W9-BRG-941

! addressing by template Pg 94

Complexity

LIFE AT THE EDGE OF CHAOS

Also by Roger Lewin

Bones of Contention

Darwin's Forgotten World

In the Age of Mankind

Origins (*with Richard Leakey*)

Origins Reconsidered (*with Richard Leakey*)

Thread of Life

Complexity

LIFE AT THE EDGE OF CHAOS

Roger Lewin

Macmillan Publishing Company
NEW YORK
Maxwell Macmillan Canada
TORONTO
Maxwell Macmillan International
NEW YORK • OXFORD • SINGAPORE • SYDNEY

Copyright © 1992 by Roger Lewin

All rights reserved. No part of this book may be reproduced or transmitted in any form or by any means, electronic or mechanical, including photocopying, recording, or by any information storage and retrieval system, without permission in writing from the Publisher.

Macmillan Publishing Company
866 Third Avenue
New York, NY 10022

Maxwell Macmillan Canada, Inc.
1200 Eglinton Avenue East, Suite 200
Don Mills, Ontario M3C 3N1

Macmillan Publishing Company is part of the Maxwell Communication Group of Companies.

Library of Congress Cataloging-in-Publication Data

Lewin, Roger.
Complexity: life at the edge of chaos/Roger Lewin.
p. cm.
Includes bibliographical references and index.
ISBN 0-02-570485-0
1. Complexity (Philosophy) 2. Chaotic behavior in systems.
3. Science—Philosophy. 4. New Age movement. I. Title.
B105.C473L48 1992 92-30314 CIP
003′.7—dc20

Macmillan books are available at special discounts for bulk purchases for sales promotions, premiums, fund-raising, or educational use. For details, contact:

Special Sales Director
Macmillan Publishing Company
866 Third Avenue
New York, NY 10022

10 9 8 7 6 5 4 3 2 1

Printed in the United States of America

For Gail

CONTENTS

Acknowledgments ix

Chapter One: The View from Chaco Canyon 1

Chapter Two: Beyond Order and Magic 23

Chapter Three: Edge of Chaos Discovered 44

boundary between chaos & inform-ation... 51

Chapter Four: Explosions and Extinctions 63

Chapter Five: Life in a Computer 84

Chapter Six: Stability and the Reality of Gaia 106

Chapter Seven: Complexity and the Reality of Progress 130

Chapter Eight: The Veil of Consciousness 150

Chapter Nine: The View from the Edge 172

A Selected Bibliography 197

Index 201

ACKNOWLEDGMENTS

The science of Complexity turned out to be one of the most intellectually stimulating sets of ideas I've come across for a very long time, and I have to thank my agent, John Brockman, for steering me in that direction. Usually, where you find interesting ideas you also find interesting people, and this experience was no exception. I should like to express my gratitude to these people, who made time in their busy schedules to talk to me and respond thoughtfully to what occasionally must have seemed like crazy questions. Effectively, the book is about them and their work, and you will find their names in the pages that follow. I owe a special debt of thanks to a few among them who gave extraordinarily generously of their time and their ideas. They are Stuart Kauffman, Chris Langton, Norman Packard, Stuart Pimm, and Tom Ray. And to Tom I also have to say: Thanks for my first experience of a rain forest. Patricia Crown, Jeff Dean, and Chip Wills organized the visit to Chaco Canyon, and introduced me to a slice of archaeology that was new to me. The memory of the event will always be treasured. The staff of the Santa Fe Institute, crucible of the new science of Complexity, were patient and helpful with my every request, for which I should like to say: Thank you all! And Trish Hoard was always resourceful in tracking down the most obscure references: Thanks.

Many were kind enough to read parts of the manuscript, and

made helpful comments. My constant supporter throughout, and my keenest editorial critic, was Gail, my wife. Without her, the process of writing would have been less rewarding and the finished product poorer.

Complexity

LIFE AT THE EDGE OF CHAOS

CHAPTER ONE

The View from Chaco Canyon

The climb was short but steep, more of a scramble really. The path had been visible from the road, twisting over rough debris at the foot of a rugged cliff. But it soon disappeared, snaking into a narrow, gloomy crevice formed when, many years ago, a huge slab of rock partially separated from the cliff face. Brown sandstone pressed in close on both sides, yielding a passageway barely wider than my shoulders, but soaring 120 feet to the cliff top. I looked up and saw a long gash of sky where the path reached its opening, dark blue against brown. I was eager to be there, but the foothold demanded close attention, sometimes smooth sand, sometimes treacherously uneven rocks. My three companions emerged first from the climb, and took the opportunity to catch their breath before I joined them.

"From up here you'll get a good idea of how it all worked," said Chip Wills. "It was a vast, complex system, nothing like it anywhere else." He swept an arm through 180 degrees by way of emphasis. We were facing south, looking across a steep-sided canyon half a mile wide. It hardly seemed possible that the small river I could see coursing irregularly along the canyon floor could have hewn so impressive a feature in the landscape. Lined by tall cottonwoods, brilliant yellow in their fall foliage, the river completed a vista of arresting beauty. This is Chaco Canyon in the San Juan Basin, New Mexico, a location that typifies the majestic scenery of the American Southwest: mesas, buttes, and canyons, stark yet softened by warm earth hues. It is also the site of some of the most important early archaeological remains north of Mexico. "Come on," said Chip, an archaeologist at the University of New Mexico and a specialist in early Southwestern cultures. "We'll get a great view of Bonito from further along here."

Almost a millennium ago Chaco Canyon was the center of Anasazi culture. It was the focus of a web of economic, political, and religious influence that encompassed more than a hundred thousand square miles of what is now harsh, unforgiving terrain, the Colorado Plateau. No other pre-Columbian society north of Mexico reached as complex a stage as this one. Archaeologists call it the Chaco phenomenon.

Chip led as we made our way along the northern rim of the canyon. Patricia Crown, of Arizona State University, and Jeffrey Dean, of the University of Arizona, also experts in Southwestern archaeology, completed the group. The arid sandstone under our feet had been laid down some 80 million years ago, when a giant inland sea divided North America into eastern and western subcontinents. Now, elevated and eroded by a conspiracy between the elements and time, a sparse desert vegetation grows in the poor soil the sandstone provides, watered sporadically by infrequent rains. Amid parched grasses, snakeweed, and other ground-hugging herbs, here and there grow stunted juniper with red leaves, sagebrush, each a cloud of feathery silver-gray leaves, and squawbush, a member of the poison ivy family, used extensively in basket making, hence its name.

Always eager to enjoy the smell of herbs, I crushed some sagebrush leaves under my nose. For the next two hours I suffered a stream of tears and a constantly dripping nose. "Distinctive pungent aroma (causes hayfever)," was a description I later read. I could confirm that. Patty related the story of a friend who, new to the region, stuffed a turkey with the leaves, thinking them to be the same as culinary sage. She did not repeat the mistake. Jeff explained that although the Anasazi had used sagebrush for many different purposes, including as material for ceremonial cigarettes and as an antidote to snake bites, no one found any use for it now. "A pity, because it's everywhere," he said.

By this time we had gone about a quarter of a mile along the rim, stopping on the way to examine two circular holes, about fourteen inches in diameter, cut into the rock. Remnants of some kind of signalling system, someone had recently speculated. Chip, ahead of us, was waving. "There," he said, pointing to the canyon floor. "That's Bonito." Patty and Jeff had seen it many times before. Their work had frequently brought them here. But familiarity did not dull the moment. Shaped like a capital D, and measuring five hundred feet on its straight side, Pueblo Bonito was the biggest of the so-called Great Houses of the Chaco culture. "Isn't that something," said Patty. We stood in silence, just looking. The early morning sun threw long shadows from the myriad walls. Someone was walking slowly in one of the open spaces, a tiny figure emphasizing the grandeur of the structure.

"The striking thing about Chacoan architecture is that the buildings pop up out of the ground, literally," said Jeff. "You can see, around that section there, some parts were four, five stories high." He was pointing to the rounded part of the D, where a multistoried catacomb of interconnecting rooms hugged the curve, five, six, sometimes seven rooms deep, pushing into the open space of the D. The straight edge, which parallels the cliff wall and faces into the canyon, was just one room deep, one story high. From the midpoint of the straight edge, a line of rooms and sunken circular structures cuts across the open space, dividing it in two. The more you looked, the more of these circular structures came into view elsewhere, about twenty-five of them, some small, some as much as fifty feet in diameter, with square and rectangular features in their base. "They are kivas," explained Patty. "They were for ceremonial gatherings, especially the big ones." She said that some of the features in the base of the kivas were structural, placements for huge wooden pillars that supported a wooden roof. Other features were symbolic, like the small, circular pit often in the middle of the floor, portal to the spirit world.

From the canyon rim, Bonito looked impressive, with its six

hundred rooms, surely the site of intense activity. Archaeologists once estimated that as many as five thousand people lived here. Close up, Bonito is just as striking as from the elevation of the canyon rim. The walls were constructed on their face from thin slabs of sandstone, producing a tightly knit, intricate pattern. Often massively thick, the walls have stood almost a thousand years without mortar, their strength derived from precise construction and sheer bulk. Long, accurate curves, invisible joins, quixotic angles for windows and doors, some characteristically T-shaped, the features of Bonito reveal the Anasazi as accomplished architects and meticulous builders. "The walls are incredibly beautiful to our eyes, but, you know, they were often faced with plaster, so you wouldn't see the details," said Patty. "That's just one of the puzzles of Chaco," said Chip. "There are many more."

As many as 70 million pieces of sandstone went into the building of Bonito, thirty thousand tons of rock that had to be hauled ten miles, shaped, and carefully slotted into flawless lines of design. More than twenty-six thousand trees were used for roof beams and posts, some of which weighed seven hundred pounds, and each imported to the site from at least fifty miles distant. "Some of the timber came from over there," said Jeff, pointing to the distant western horizon, the Chuska Mountains, where ponderosa pine, Douglas fir, and spruce still grow. An expert in tree-ring dating techniques, Jeff is more familiar than anyone with the wood in the Chaco Great Houses. He has taken cores from many of the beams, producing an extensive catalogue of dates throughout each of the pueblos. "You can follow the sequence of construction very closely, using the dates," he explained. He also said that pretty soon it may be possible to pinpoint the geographical source of each roof beam, by matching trace element patterns in them with those in modern forests of the region.

Twenty-six thousand trees, carried fifty miles, with no transport but human muscle power and ingenuity: such numbers demanded some back-of-the-envelope calculations. Assuming six people per tree, and a four-day round trip, I came up with more than seventeen hundred person-years of labor. And Bonito was just one of nine such

Great Houses here, six on the canyon floor and three at various positions on the mesas. "That's what impresses me about all this," said Chip. "You don't get the feeling of a people scratching a living. You sense an exuberance, a people capable of organizing tremendous feats of construction, including irrigation and farming under challenging circumstances. No doubt about it, Chaco was an important place, very important." Discoveries of turquoise jewelry and a few high-status burials at Bonito speak of ancient ceremonies.

Chaco is also important to today's New Agers, who flock to the canyon for their own ceremonies, complete with borrowed Buddhist chants, meditation techniques, and crystals. "They come to Casa Rinconada, just over there," said Jeff, pointing to a slightly raised area on the other side of the canyon. One of three isolated Great Kivas, each strategically situated on the canyon floor, Rinconada measures sixty feet in diameter, is entered via a tunnel of symbolic structure, and has niches cut into its circular wall. It is easy to imagine the power of such a place: roofed by wooden poles, its gloom brought to life by flaming torches, sacred objects in the niches, the steady chanting of religious leaders. "That's why the New Agers love it," said Jeff. To them, and the ancient Anasazi, the Great Kivas are sacred places.

Jeff told us of a New Age group that had visited Rinconada a year earlier, from somewhere in Texas. As they were leaving, one of them died of a heart attack as he was getting in his car for the journey home. His friends took him home, had him cremated, and then returned with his ashes, which they scattered over the floor of Rinconada. "The Navajo were appalled," said Jeff. "They can't abide the sense of death in the kivas." The floor of the kiva had to be scraped clean of mortal contamination before the Navajo would return. "A lot of the archaeological detail was destroyed in the process."

The impression of Chaco as a place of importance to the Anasazi was enhanced as we turned away from the canyon rim, and headed north toward the remains of another Great House, Pueblo Alto. With some difficulty we located the remnants of a road, constructed almost a thousand years ago by the Anasazi to link Bonito and Alto.

Sometimes running over relatively level terrain, sometimes ascending steep inclines by means of steps cut into rock, the road held to a straight line between the two settlements. But the road was odd. With no wheeled transport and no horses, the Anasazi would have made it easier for themselves had they followed the natural contours of the land rather than challenging them at every step. A simple trail would have sufficed, not a vector some thirty feet wide. For the Anasazi, their roads, like their architecture, surely went beyond the merely functional.

Anasazi roads have been known since the beginning of the century, but only from scattered fragments. Only when remote sensing techniques became available in the 1970s did their nature and extent become apparent. "They run straight, they radiate out of Chaco Canyon, and they run a long way," said Jeff. "But they don't go in every direction. There's not much to the east, for instance." The roads may follow earlier routes that were used for the transport of supplies to Chaco, Jeff speculated. If this was their origin, they later assumed another role. The roads are now known to connect outlying settlements, some a hundred miles distant, to Chaco Canyon. The settlements, between 150 and 300 of them, are Chacoan in their architecture and organization, and clearly formed a unified social system of some kind.

Soon we reached Pueblo Alto, smaller than Bonito and less extensively excavated. We turned and looked back, straining to see where the road we had followed reached the rim of the canyon, before it descends as a steep staircase cut into the cliff face. Beyond, on the opposite side of the canyon, is South Gap, an exit from the canyon through which runs a major road. In the far distance Hosta Butte stands like a sentinel in the desert. "The approach to the canyon on that road was clearly meant to be impressive," said Chip. "It's extensively engineered, at least thirty feet wide, and as you enter the canyon banks are built up on either side of the road, higher and higher. You'd get the impression of gradually sinking, like Chaco was swallowing you up."

I was forming an idea of how it all may have worked, as Chip had promised. The nine Great Houses, and several isolated Great Kivas of Chaco Canyon, were not only the geographical center of the Anasazi a millennium ago. They were in some way the center of powerful influence. I assumed that, by analogy with modern states, Chaco represented a capital of some kind, perhaps with Pueblo Bonito the principal center. "Recent excavations at Pueblo Alto, and reassessment of Bonito, shows that whatever the Chaco phenomenon was, it wasn't something we are familiar with today," said Chip, breaking my thought. "The Great Houses weren't densely populated; maybe only a few hundred people lived at Bonito, for instance. Certainly not the thousands we used to imagine." Gone, the notion of Pueblo Bonito with its bustling population of five thousand. Now, the image of a few score people, some helping to fulfill the numinous function of this architecturally elaborate and beautiful place while others tended the fields, where corn was farmed.

Some archaeologists have speculated that the many rooms in the Great Houses were for storage, so that Chaco would have been a giant distribution center. But there is little direct evidence for that idea, and the intricate configuration of many of the Great Houses, and the prevalence of kivas, argues against it. "Many archaeologists think about it as a ceremonial center," said Patty. "A few people lived here, some perhaps as caretakers, others as important figures in the ceremonies. But most came as visitors, perhaps periodically for seasonal rituals." There's no doubt that stone and pottery were brought to Chaco in great quantities and from distant parts, and seashells and turquoise, too. "See that mound over there," said Patty, pointing to rough terrain to the east of Pueblo Alto. "That's a trash heap, full of broken pottery."

Sure enough, we could see scores of potsherds, some plain, some decorated. Intact, these pots would have born the characteristic patterns of the time, some made like rope baskets, others with black-on-white geometric decoration. A comfortable size to cradle in two hands, they were objects of beauty as well as utility. When the twelve-foot-high mound was excavated a decade and a half ago, an astonishingly large number of pot fragments were found, particularly for so modest a settlement as Pueblo Alto. About twenty-five hundred vessels a year had been broken, which works out at twenty-five pots per person every year. "Either they were pretty clumsy or they weren't using pots in a conventional way," remarked Chip. "The excavation showed that the sherds were dumped periodically, not as you'd expect from daily use. Maybe this just tells you Alto was occupied seasonally. Or maybe there were ceremonial events that involved breaking pots. My guess is the latter."

As we wandered back to the ruins of Pueblo Alto, Chip remarked on the sense of elevation, of dominance of this spot, with horizon-to-horizon vistas. Only two other Chaco Canyon Great Houses enjoy a 360-degree view of the countryside. Close to the Continental Divide as it is, this region is high, some seven thousand feet. From the north, four or five Anasazi roads converge on Pueblo Alto, coursing arrowlike across the high plateau. A swing of the gaze from north through west, south, and a little to the east, encompasses the vast terrain under the influence of the Great Houses of Chaco Canyon.

Lunchtime had arrived and we found a sheltered spot toward the western end of the site, most of its walls rounded by centuries of erosion. The recent excavation had uncovered about 10 percent of the structure, revealing variants on the Chacoan stonework. Chip indicated a small room nearby. "Three grinding stones were found in there, lined up side by side," he said. "In front of each of them, impressed into the clay floor, was the shape of a basket. You could imagine three people methodically grinding corn, periodically depositing the flour into the baskets, gossiping about people and life at Pueblo Alto." So ordinary an activity in so special a place.

The sun was at its fall high, the sky a hazy blue, signalling tomorrow's forecast storm. Despite a steady breeze we were warm. We quietly enjoyed the tranquillity of the place, ruined walls holding ancient secrets.

Santa Fe seemed a long way away.

Jeff, Patty, Chip, and I had planned our trip a year earlier, at the end of a scientific conference in Santa Fe, which is 120 miles almost due east of Chaco, nestled in the saddle between the Jemez and Sangre de Cristo Mountains. The title of the conference, "Organization and Evolution of Southwestern Prehistoric Societies," was ordinary enough. Many of the participants were anthropologists and archaeologists, as would be expected at such a gathering. But there were physicists, too, computer experts, and a theoretical biologist. One of the organizers was Murray Gell-Mann, a Nobel Prize winner from the California Institute of Technology, who is better known for discovering mysteries of the quark than for uncovering past civilizations. The gathering was under the auspices of the Santa Fe Institute, where, as the *Wall Street Journal* recently observed, "no idea is too crazy."

The archaeologists were there to try to understand more of the larger pattern of Southwest prehistory. Why, for instance, did the introduction of maize agriculture into the region three thousand years ago initially have so little impact on social organization? Similarly with ceramics a little more than a thousand years later. What prompted the burst of new forms of social organization from A.D. 200 onward? What was behind the rapid rise of Chaco Canyon as an important regional center between A.D. 900 and 1150, the Chaco phenomenon? Equally, why did Chaco collapse, never to regain the status it once enjoyed?

Chaco never reached the level of social complexity that can be called a city-state, such as had arisen earlier in Mexico, Central and South America, and in the Old World. But unquestionably it included elements of social and economic organization that are precursors to state formation, a subject that has long enthralled prehistorians. The archaeologists and anthropologists who attended the Santa Fe meeting therefore had an opportunity to think about the larger picture of state formation, and to analyze some of its details.

For the folks of the Santa Fe Institute itself, the motive was different. For them, cultural evolution and state formation may be yet another example of an important general phenomenon. Since it was established in 1984, the institute has attracted a core group of physicists, mathematicians, and computer whizzes. The computer is their microscope, through which they scrutinize the world, real and abstract. No corner of the natural world escapes their gaze:

chemistry, physics, biology, psychology, economics, linguistics, human society, all are encompassed in a common intellectual orbit. Unnatural worlds are included, too, worlds of the most tenuous existence created in computers. The phenomenon that may link these disparate worlds, including what propelled Chaco Canyon along its unique history, is called complexity.

To some, the study of complexity represents nothing less than a major revolution in science. Among them was Heinz Pagels, whose stellar career as a physicist at Rockefeller University was tragically ended in 1988 by a climbing accident. "Science has explored the microcosmos and the macrocosmos. . . . The great unexplored frontier is complexity," he wrote in *The Dreams of Reason*, published the year he died. "I am convinced that the nations and people who master the new science of Complexity will become the economic, cultural, and political superpowers of the next century." That's quite a claim for a science that as yet has perhaps a few dozen active practitioners, a science that most people have never heard of, and if they have, ask: Is that the same as chaos?

I've asked many people the same question. "Chaos and complexity are chasing each other around in a circle trying to find out if they are the same or different," responded Chris Langton. Chris, who is a member of the institute and participated in the Southwest Prehistory conference, is one of those people who find it difficult to talk without at the same time dashing to the blackboard to illustrate what he's saying. "Totally ordered over here. . . . Totally random over here," he sketched, with broad strokes. "Complexity happens somewhere in between." Stab. We were talking during a break in the conference, and I wanted to come to grips with what seemed to me mercurial concepts. Why don't you find complexity over there, in the random area? I asked. That seems pretty complex to me, hard to describe. "It's a question of structure, of organization," he said. "The gas in this room is a chaotic system, very random, very little order. The science of Complexity has to do with structure and order."

Order, such as you see in the social organization of Chaco? "Yes." Order, like when an embryo develops to become a fully formed adult? "That too." What about patterns of evolution? "Yep." And ecosystems? "Absolutely." If that's so, I wondered why the institute isn't full of anthropologists and biologists rather than physicists and computer jocks. "Because we're looking for the fundamental rules that underlie all these systems, not just the details of any one of them," Chris explained. "You can only understand complex systems using computers, because they are highly nonlinear and are beyond standard mathematical analysis." And, he said, so far few biologists are aware of complexity as it is understood at the Santa Fe Institute. "If they were they'd probably think we're nuts."

For three centuries science has successfully uncovered many of the workings of the universe, armed with the mathematics of Newton and Leibniz. It was essentially a clockwork world, one characterized by repetition and predictability. The launching of a spacecraft to rendezvous with the Moon after several days of travel depends on that predictability. Alter the trajectory of the spacecraft just slightly, and its new path, which deviates just slightly from the original, can again be predicted using the equations of motion. That's a linear world, and it is a very important part of our existence. Most of nature, however, is nonlinear and is not easily predicted. Weather is the classic example: many components interacting in complex ways, leading to notorious unpredictability. Ecosystems, for instance, economic entities, developing embryos, and the brain—each is an example of complex dynamics that defy mathematical analysis or simulation.

In nonlinear systems small inputs can lead to dramatically large consequences. This is often characterized as the so-called butterfly effect: a butterfly flaps its wings over the Amazon rain forest, and sets in motion events that lead to a storm over Chicago. The next time the butterfly flaps its wings, however, nothing of meteorological consequence happens. This is a second feature of nonlinear systems: very slight differences in initial conditions produce very different outcomes. That's the basis of their unpredictability. If the laws of motion were nonlinear, no sane astronaut would be willing to be blasted off to the Moon, because the chance that the ground crew would be able to arrange initial conditions—weight, altitude, acceleration, and the like—precisely enough to determine the outcome would be minuscule. Almost certainly he or she would finish up anywhere but the Moon. Baseball players rely on the linear—that is, predictable—nature of the laws of motion, too; otherwise fielders would be unable to position themselves for the spectacular catches they make.

Classical physics regarded complex systems as exactly that: systems that, when powerful enough analytical tools were eventually at hand, would require complex descriptions. The central discovery of the recent interest in nonlinear dynamical systems is that this assumption is incorrect. Such systems may indeed appear complex on the surface, but they may be generated by a relatively simple set of subprocesses. The discovery of chaos theory was in the forefront of that emerging understanding of nonlinear dynamical systems, as James Gleick so enthrallingly described in his book *Chaos*. Many of the people who, against the better judgement of their more experienced colleagues, pursued an understanding of chaos are now involved with the wider issue of complexity. Still viewed askance by some, they are no longer regarded as completely misguided.

I asked Chris if it was fair to say that chaos is a subset of complexity. "Yes it is, in that you are dealing with nonlinear dynamical systems," he replied. "In one case you may have a few things interacting, producing tremendously divergent behavior. That's what you'd call deterministic chaos. It looks random, but it's not, because it's the result of equations you can specify, often quite simple equations. In another case interactions in a dynamical system give you an emergent global order, with a whole set of fascinating properties." Chris is at the board again, rapidly sketching a cluster of small circles, joined by double-headed arrows. "These are the components of your system, interacting locally." Above them appears what looks like a child's version of a cloud, and a volley of large arrows shoots up from the cluster below. He then added two arrows, one emerging from each side of the cloud, sweeping down toward the cluster. "From the interaction of the individual components *down here* emerges some kind of global property *up here*, something you couldn't have predicted from what you know of the component parts," continued Chris. "And the global property, this emergent behavior, feeds back to influence the behavior of the individuals *down here* that produced it."

Order arising out of a complex dynamical system, was how Chris described it, global properties flowing from aggregate behavior of individuals. For an ecosystem, the interaction of species within the community might confer a degree of stability on it; for instance, a resistance to the ravages of a hurricane, or invasion by an alien species. Stability in this context would be an emergent property.

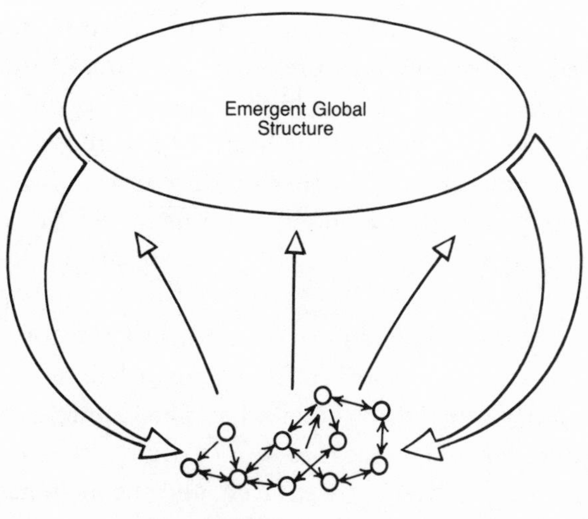

Fig. 1. Chris Langton's view of emergence in complex systems.

In industrial societies, the aggregate behavior of companies, consumers, and financial markets produces the modern capitalist economy, "as if guided by an invisible hand," as the Scottish economist Adam Smith once put it. For a growing embryo, the global consequence of the aggregate of a kaleidoscope of developmental processes is a mature individual. And for the brain, billions of neurons interact to yield complex behavior patterns. Including consciousness? I asked Chris. Could the theory of complex systems explain consciousness? "If the theory of complex systems is not some kind of seductive illusion; and if the brain can be described as a complex adaptive system; then, yes, consciousness can be explained, too," said Chris. "At least in principle."

By now it was clear to me that there is tremendous scope for confusion over terms, like *chaos* and *complexity*. For most of us *chaos* means random. In the realm of nonlinear dynamical systems research, this is not the case. And for most of us, too, *complex* can mean almost the same as chaotic. As Chris said, the molecules in the room I'm now sitting in as I write this chapter are maximally chaotic, and to describe them would require the documentation of the position and activity of every one of them. No simpler description is possible. By some measures that would make the room full of molecules very complex. That kind of complexity did not interest me, nor is it the main focus of the Santa Fe Institute folk. They are interested in complex systems that produce order, as I am.

Murray Gell-Mann has a good phrase for it: Surface complexity arising out of deep simplicity.

The thirty or so participants in the Southwest Prehistory conference assembled on the first morning in the main hall of the Santa Fe Institute, clearly a little unsure of what would unfold. With what some institute members took as a nice irony, the institute was housed in a former convent, Cristo Rey, and the main hall had been the chapel. A low adobe structure situated at the end of Canyon Road, well known these days for its art galleries, the convent building provided the kind of intimate atmosphere necessary for nurturing tentative ideas. This was just the kind of milieu that the institute's founding members had sought, explained Gell-Mann in a brief introductory session. "Away from universities or established institutions, where bureaucracies and academic barriers restrict creative thought," he said.

The notion of an interdisciplinary, unrestricted institute emerged in 1983 from informal lunchtime discussions among a group of fellows at the nearby Los Alamos National Laboratory, including Gell-Mann, a frequent visitor. Best known as the birthplace of the atomic bomb, Los Alamos also boasts deep expertise in nonlinear systems analysis. During those lunchtime discussions there developed an inchoate sense that something of broad importance could emerge from such analysis, if only it were allowed freer rein. No discipline would be excluded; all would be included under the umbrella of complex systems. In particular, as it developed with the maturation of the institute, complex adaptive systems.

"Turbulent flow in a liquid is a complex system," explained Gell-Mann in the introductory session. "But it can't be called adaptive. In turbulent flow there are eddies that give rise to smaller eddies and so on, and certain eddies have properties that allow them to survive in the flow and have offspring, while others die out. There's information in the system, no question. But it doesn't produce a schema, a compression of information with which it can predict the environment."

Gell-Mann is a man of not uncommon intensity but of distinctly uncommon abilities. Along with his Nobel prize in physics he lists linguistics as a serious study (he speaks thirteen languages, each with perfect locution), and psychology, anthropology, archaeology, ornithology, and cultural and ecological conservation, too. He will tell you he takes on fifty times more than anyone else can do, and as a result falls eight years behind with each passing day. He claims he works at only 2 percent efficiency, though it is difficult to believe. Gell-Mann has little to be modest about, and isn't.

"In biological evolution, experience of the past is compressed in the genetic message encoded in DNA," Gell-Mann continued. "In the case of human societies, the schemata are institutions, customs, traditions, and myths. They are, in effect, kinds of cultural DNA." Complex adaptive systems are pattern seekers, Gell-Mann said. They interact with the environment, "learn" from the experience, and adapt as a result. The notion that complex adaptive systems encrypt information about their environment, know their environment in some special sense, was appealing. A genetic package as encryption of the environment in the context of biological evolution? Of course. And cultural institutions for human societies? That too. But what of other systems, like ecosystems, like embryos, how might they know the world they inhabit? That, and other such questions would have to wait. With Gell-Mann's final comment—a warning really—that "some schema may be maladaptive" still fresh in our minds, the plenary session broke up and participants sorted themselves into previously allotted groups, each with a particular task to address.

Patty, Chip, and Jeff were the archaeologists in my group, and we were joined by Chris, and Stuart Kauffman, a theoretical biologist from the University of Pennsylvania, also with ties to the institute. We assembled in Stu's office, just big enough for the six of us. My familiarity with the archaeology of the Southwest was limited at best. Most of my forays into prehistory had been in the deep past, such as Olduvai Gorge and Koobi Fora, in East Africa. I'm used to thinking about millions of years in the past, not a few thousand or less. A powerful intuition had brought me to these novel surroundings, an intuition that I would get a glimpse of a pattern, a first glimpse that would lead me on a journey of discovery. Patterns in nature grasp my interest, patterns in evolution, in ecology, in the history of life on Earth. Gell-Mann's phrases, "Surface complexity arising out of deep simplicity" and "Complex adaptive systems are pattern seekers" resonated with that intellectual quest. And I wondered, as I sat in that small room, how clear the pattern might be.

"You've run into the best practitioners of a very questionable process," said Stu, indicating himself and Chris. "It might give you something to think about. It might not." It was obvious which he thought the most likely. Some archaeologists view history as having no direction at all, "just one damn thing after another," so to speak. In which case, Chip asked, how could Stu's "questionable process" lead to a better understanding of the process? What followed was a probing exchange of ideas, questions, and responses, pure archaeology interleaved with abstract complexity analysis and biological analogies. Frank bafflement was not uncommon, and at first it was difficult to see where the discussion might lead. Gradually, points of contact began to crystallize, tentative insights accreting on initially tiny possibilities.

There was talk of how avalanches of goods in and out of an economy can be analyzed as a result of an innovation, such as the automobile. Could this be used in an archaeological setting? Chip said it was possible to see some of this in the Southwest, with the introduction of corn in 1000 B.C., and then pottery later. But the puzzle was that very little else changed, for more than a thousand years. Corn remained a minor part of subsistence. What was missing during that time? Social organization? Means of accumulating surpluses? "At A.D. 200 pottery became important, irrigation started, sedentism, too, more complex social organization," said Patty. "Something happened to produce a big change. And it happened fast."

"That's my phase transition," exclaimed Chris, leaping to the board, where he quickly sketched the classic physics textbook diagram of a phase transition, with solid, fluid, and gaseous phases. "As you approach the boundary and cross it, you suddenly get a phase change," Chris explained. "Here, you're in one phase, there another, and it switches very quickly, pushed by a slight change in conditions, temperature and pressure in this case." Maybe some small change in external conditions pushed the Anasazi in A.D. 300 to 400 from a simple foraging band structure to something more complex, Chris speculated, perhaps a specialization of tasks. At this point, Norman Yoffee joined the group as a brief visitor. An anthropologist at the University of Arizona and an expert in the dynamics of state formation, Norman described the history of early civilizations in Mesopotamia, modern Iraq. "When you see state formation, it always happens quickly," he said. "States are expectable and predictable."

Chris quickly reiterated what he'd said about phase transitions in physics and their analogy to other systems, including the shifts between different levels of social complexity. "I see everything through phase transition glasses," he conceded. He offered another example, that of the change from single-celled organisms to multicellular organisms, which occurred 600 million years ago, the Cambrian era. For 3 billion years, from soon after the Earth cooled sufficiently, the highest form of life was the single cell. True, a degree of complexity emerged a little more than a billion years ago, when cells developed packaged nuclei and included mitochondria. Nevertheless, it was eon upon eon of mind-numbing sameness. Then, suddenly and with spectacular effect, the trick of cellular differentiation and aggregation into multicellular organisms evolved. An explosion of new forms occurred, with a bewildering variety of complexity.

"Cellular specialization happened in the Cambrian, and . . . Bang! . . . all hell broke loose," said Chris graphically. "How about that for an analogy for what happened in the Southwest?" he asked. But Chris had in mind something more than simple analogy, something more than mere coincidence of pattern. "Maybe there's something fundamentally the same about the two systems, so that the patterns are the same, no matter what the details of the system are," he speculated.

Biologists know the onset of multicellularity and the subsequent burgeoning of complex forms as the Cambrian explosion, a massive punctuation in the history of life. In anthropomorphic terms, it seemed to have been a time of unrestrained evolutionary experimentation, a time when any and every possible body plan was tried. Many forms seem to have gone extinct in short order (that is, within 50 to 100 million years), leaving a diminished range of body plans, or phyla, from which modern organisms of all forms are built. Chip, Patty, and Jeff were intrigued with the analogy—call it that for the moment—between the Cambrian explosion and bursts of social change in the history of the Southwest, and gave examples of similar patterns in the history of the Anasazi. I knew that George Gumerman, the co-organizer of the conference, also considered the analogy worth exploring.

In a general introductory paper to the conference, George had referred to the coincidence of patterns, biological and cultural. "The increase in variety of social conventions is in many ways analogous to the 'Cambrian Explosion,' " he wrote. "The sudden increase and richness of life forms in the Cambrian has been attributed to the occupation of a 'vacant ecology,' an environment which was available for and receptive to evolutionary experimentation. Across the

Southwest there was such an increase in population and range expansion that almost every niche was filled by A.D. 1100. Furthermore, numerous types of social organization were experimented with and succeeded for a short time." The Chaco Canyon tradition was one of those experimental organizations, and it was among the most complex and successful.

How did you hit on the analogy? I asked George when I telephoned him shortly before the conference. He said that some months earlier he happened to be reading an issue of *Science*, the 20 October 1989 issue, which contained an article by Cambridge University geologist Simon Conway Morris, on the Cambrian explosion: "Burgess Shale Faunas and the Cambrian Explosion," was the title of the paper. "I didn't follow it all, because I'm not a paleontologist," George told me, "but I could see the overall pattern, and thought, That's just like the pattern we see in the Southwest. I'd say it's a nice analogy. Maybe it's universal to all evolving systems, maybe to all complex systems. I don't know." This explicit analogy had done much to encourage my intuition that many patterns in nature were in some ways variations upon similar themes. It had brought me to Santa Fe.

"The pattern of the Cambrian explosion is fundamental to all innovation," said Stu, responding to the group's interest in this line of argument. "You get an initial scatter of new forms, and then it gets harder and harder to improve on them. You see it in biology. You see it in industrial economies." And, maybe, you see it in the evolution of social complexity.

The evolution of social complexity has taxed anthropologists for many years, in terms of definition and explanation. No one doubts that a nation state is more complex than a foraging band. But the process of moving from one to the other, and the nature of the steps on the way, still causes debate. Three decades ago, University of California, Santa Barbara, anthropologist Elman Service proposed a neat structure, which progressed from foraging band, to tribe, to chiefdom, to state, a predictable evolution of particular forms. Too neat, as it turns out. There are many local variations that make the classification, and step-by-step progression, seem simplistic. Chaco, for instance, doesn't fit well into the once-accepted definition of chiefdom, lacking many of the trappings of such a power structure, such as monumental architecture and elaborate burials. And yet there is no question that Chaco represents a significant increase in social complexity over the foraging band or simple agricultural village.

"If cultural evolution works like other models of evolving systems, you'd expect to see rapid transitions from one level of organization to the next," continued Stu. "We've talked about the pattern of innovation at the point of change. Is it reasonable to look at the different levels as something special, whether you call them tribes, chiefdoms, or something else?" Not surprisingly, the archaeologists in the group found many ways to answer that challenge, explaining why some people think one way, others another. "I don't think it's completely unreasonable," Jeff eventually conceded. "There might be discrete levels of organization that in a general sense are common to all of cultural evolution. But don't ask me what you should call them." He explained that some archaeologists refer to the shift from one level to another as hinge points. Evolutionary biologists would call them punctuations. And physicists, like Chris, would say phase transitions.

If there really are discrete levels of social organizations common to all of cultural evolution—no matter what you call them—I wondered whether each level represents some kind of "natural" system, a level of organization to which the evolving cultural system is inexorably drawn. The archaeologists were properly skeptical. But, again, Jeff allowed that it was not unreasonable, or at least no one could unequivocally prove it wrong. I realized I was pushing the envelope of what was reasonable and knowable. But that was why I was there.

Most complex systems exhibit what mathematicians call attractors, states to which the system eventually settles, depending on the properties of the system. Imagine floating in a rough and dangerous sea, one swirling around rocks and inlets. Whirlpools become established, depending on the topography of the seabed and the flow of water. Eventually, you will be drawn into one of these vortexes.

There you stay until some major perturbation, or change in the flow of water, pushes you out, only to be sucked into another. This, crudely, is how one might view a dynamical system with multiple attractors: such as cultural evolution, with attractors equivalent to bands, tribes, chiefdoms, and states. This mythical sea would have to be arranged so that the hapless floater would be susceptible to whirlpool one first, from which the next available would be whirlpool two, and so on. There would be no necessary progression from one to two to three to four. History is full of examples of social groups achieving a higher level of organization, and then falling back. This is what happened at Chaco. And, until recent times, every society that has achieved level four—the state—has eventually collapsed.

Pushing this line of thought to its limit, I asked Chris what he would expect if he were able to build a computer model of cultural evolution. It would have to begin with the components of the foraging band, and its social and economic dynamics. Would he expect the model to exhibit attractors equivalent to discrete levels of organization, like the tribe, the chiefdom, the state? "I'd expect to see attractors, definitely," Chris said without hesitation. "If you have populations that interact, and their fitness depends on that interaction, you will see periods of stasis punctuated by periods of change. We see that in some of our evolutionary models, so I'd expect to see it here, too." In which case, history couldn't be described as one damn thing after another, could it?

With heady stuff like this emerging it was clearly time to break up the group and drink some tea in the courtyard of the convent.

"I've seen you folks before," said Chris. "You weren't archaeologists. You were biologists. You were linguists. You were economists, physicists, all kinds of disciplines." It was the final session of the conference, everyone back together in the chapel. Each group had given a summary of its discussions and conclusions. Chris was giving the Santa Fe Institute's perspective. "Each time a group of people comes here for one of these conferences, there's some kind of historical process under consideration. Evolutionary systems are like that. They're unique processes, so you can't compare them directly with anything. You'd like to rerun the process, see what happens a second time around, and a third, and so on. You can't, so that's where we come in."

Chris and others like him at the institute are looking for universal principles, fundamental rules that shape all complex adaptive systems. "You know, that conference on Southwest societies made things very clear to me," Chris told me later. "It was then I realized I'd seen the same patterns over and over again at institute conferences. That's why I said, 'I've seen you folks before. . . .' " Chris and I were talking a year after the conference, just after my return from the visit to Chaco Canyon. We were in the institute's new home, an office building on Old Pecos Trail it shares with lawyers and an insurance company. More convenient, perhaps, but it wasn't the convent. "Here are all these hunting bands out there, groups of individuals, each able to do all the tasks in the group. Each of them can hunt, gather plant foods, make clothes, and so on. They interact with each other, you get specialization, then . . . Bang! . . . phase transition . . . it all changes. You have a new level of social organization, a higher level of complexity."

Chris was at the board, busily sketching, first the band-to-tribe transition, then showing how it's the same as going from single-celled organisms to multicellular organisms. Soon he's talking about the Cambrian explosion and punctuated equilibrium. More sketching. Then the output from a simple evolutionary model run on a computer. Sketch. Before long he's on to bicycles, weeds, and the fall of Gorbachev. . . . Sketch . . . sketch.

It was then I was really sure that my journey in search of patterns was going to be worthwhile.

CHAPTER TWO

Beyond Order and Magic

"They looked at me as if I was crazy," recalled Stuart Kauffman. "There I was, shuffling this stack of cards, then handing them to the programmer." This was way back in 1965, when you fed your program and data into a computer on a set of punched cards. "Steam age." If the program was to work, then the cards had to be in perfect order, everyone knew that. One card out of place and the machine was likely to spew out garbage. "And there was I, shuffling my data cards, randomizing them. No wonder they looked at me with a kind of 'crazy kid' smirk on their faces."

Stu was twenty-four at the time, a second-year medical student at the University of California, San Francisco. His foray to the medical school's computer center had nothing to do with his allotted studies, however. He was there to prove that he was right and that the entire biological community, from Darwin on down, was wrong. "Not a modest venture," Stu admitted, as we talked in his cluttered office in the biochemistry department of the University of Pennsylvania. "But, you know, I had this unshakable conviction that I was right."

I first met Stu almost twenty years ago, at a scientific conference on circadian rhythms, in Berlin. The conference had been a mix of basic biology and weird—to me—mathematics, just the kind of intellectual combination Stu and his friends relish. He's nothing if not driven. "The highest mouth-to-brain ratio of anyone I know," a close friend and colleague of Stu's once told me. "And we're talking Big Brain." The MacArthur Foundation apparently agreed, and awarded him one of its prestigious "genius" fellowships in 1987. So when Stu called early in 1990 and said, "Why don't you come up to Penn, and I'll tell you about complexity: it's new and it's going to be real big," I knew I had to go. It was to be my introduction to Complexity as a science in its own right. This was the spring before the Santa Fe meeting on Southwest prehistoric societies.

Talking fast and with a characteristic mix of jargon and lucid metaphor, Stu explained how an inner certainty had driven him to that bizarre computer experiment as a medical student, a certainty that the conventional explanation for the origins of order in the world of nature just had to be wrong. There is order everywhere, in the morphological similarities among groups of organisms, and in the remarkable way in which individual organisms operate within their environments, as if they have been carefully designed. Scholars have been fascinated with the phenomenon since Aristotle's time. And in the mid-eighteenth century the Swedish scholar Carolus Linnaeus grouped known organisms according to similarities they displayed, his *Systema Naturae*, or System of Nature, a classification that biologists still use today. The conventional explanation of all this order is natural selection, which biologists from Darwin onward considered the force that fits organisms to their niches in the world. The similarities among groups are the result of common descent, "descent with modification" as Darwin described it.

"Natural selection was said to be the sole source of this order, an all-powerful force that could produce more or less any kind of biological form, given the right circumstances," said Stu. "Don't ask me how, but I knew that couldn't be right, that there had to be a whole lot of spontaneous order out there." Spontaneous order? That smacks of vitalism, I ventured, a once popular but now discredited notion that much of the wonder of the natural world is the consequence of an *élan vital*, or vital spirit. Nature wasn't so much explained as explained away by this notion, and it is anathema to modern science. "I don't mean that, of course," responded Stu. "I mean that self-organization is a natural property of complex genetic systems. There is 'order for free' out there, a spontaneous crystallization of order out of complex systems, with no need for natural selection or any other external force. I was sure of that back in medical school, and I still am."

Stu is still unable to account fully for his long-held conviction that the conventional wisdom must be wrong. But his decidedly unconventional education surely contributed. He'd gone up to Dartmouth in the fall of 1957, with the intention of becoming a playwright, not just an ordinary playwright but a Great Playwright. "No point in being an average playwright." Three weeks and two mediocre plays later, and prompted by a friend who had gone up to Harvard at the same time, Stu decided that he would be a Great Philosopher instead. "Kids go into philosophy because they're interested in ethics, the mind, those kind of good things," he explained.

A bachelor's degree in philosophy at Dartmouth successfully completed (Phi Beta Kappa), Stu was awarded a Marshall scholarship, which took him to Oxford University in the fall of 1961, where he read philosophy, psychology, and physiology at Magdalen College, one of the more prestigious courses at one of the more prestigious colleges at that ancient university. "A wonderful time," he declared, throwing his arms behind his neck, slipping down further into his creaking chair, and stretching his legs, memories flooding back. "Climbing the walls back into college at night, that kind of thing." What about becoming a Great Philosopher? I asked "Well, listen to this. I went through the following syllogism: 'In order to be worthwhile a philosopher you had to be as smart as Immanuel Kant. I am not as smart as Immanuel Kant. Therefore I will become a doctor.' With that kind of reasoning, no wonder I wasn't a Great Philosopher!"

Although he judged himself to be only adequate in philosophy, Stu discovered a facility for inventing theories to explain whatever challenge he was presented in psychology, including aspects of neural networks. He was so good at it that he mistrusted the gift as a certain glibness, and decided to ground himself in some solid learning. Medical school would provide the necessary grinding through facts he sought to impose upon himself. First, however, a year of premedical education was necessary, a special course at Berkeley designed to teach some background biology, including embryology. This was 1964, when Berkeley was afire—literally—with radicalism. "I wasn't much involved in demonstrations, but later on, in my third year of medical school, I did sign a declaration that I would not serve in Vietnam. Soon afterward I saw a flotilla sailing into the bay, an aircraft carrier, cruisers, and so on, and thought, 'Boy, it's going to take a lot of signatures to stop this thing.' "

The early 1960s were also a special time for molecular biology, because in the previous few years two French researchers, François Jacob and Jacques Monod, had made breakthroughs in understanding the regulation of gene activity. They discovered the existence of feedback mechanisms by which genes were switched on and off, a kind of servo system at the level of molecules, and also analogous to the binary switch system of digital computers. The work was soon recognized by the Nobel Prize committee. "It was tremendously exciting to learn this stuff," said Stu. "I became obsessed with embryology, particularly how embryonic cells differentiate, forming muscle cells, nerve cells, cells of connective tissue, and so on; how the one hundred thousand genes in the human genetic package might produce this bewildering assembly of different cell types, some 250 in all. Everything was coming into place, the Jacob/Monod ideas, even the networks I'd played around with in Oxford." Embryology, the way a single, fertilized cell multiplies, differentiates, and assembles into an adult organism, was and still is one of biology's greatest challenges. The young Kauffman was ready for the challenge, equipped with the sparsest of biological grounding and only a rudimentary grasp of the mathematical tools he planned to apply.

"My ignorance was a strength," he told me in all seriousness. "If I'd had a proper biological education and knew the math, I would have known what I wanted to do just wouldn't work. I wouldn't have tried it." Stu reasoned that it was all but impossible for natural selection to orchestrate the activity of the one hundred thousand genes in the human genome so as to generate the range of some 250 different cell types. That represents 250 different patterns of gene activity. "Do you know how many potential activity states there are in an assembly of one hundred thousand genes?" he asked, without waiting for an answer. "$10^{30,000}$. That's hugely larger than the number of hydrogen atoms in the universe. Some people will argue that natural selection will successfully lead you through the swamps of all $10^{30,000}$ states in the system, eventually hitting the 250 that you want. But I had a different solution, unthinkable and absolutely counterintuitive.

"Imagine that the genes are arranged as a network, each either active or inactive depending on the inputs from other genes," Stu began. Sounds like a parallel processing network, I said. "That's right. But, imagine that the links between the genes are randomly assigned. Would you expect to get order out of that?" Naïve about these things, I nevertheless guessed it unlikely. Stu should have thought it unlikely, too, had he been aware as a young medical student that several big names in mathematics and computation had earlier experimented with similar systems, and had found nothing interesting. "The counterintuitive result is that you do get order, and in a most remarkable way."

Systems of this sort are known as random Boolean networks, after George Boole, the English inventor of an algebraic approach to mathematical logic. The network proceeds through a series of so-called states. At a given instant, each element in the network examines the signals arriving from the links with other elements, and then is active or inactive, according to its rules for reacting to the signals. The network then proceeds to the next state, whereupon the process repeats itself. And so on. Under certain circumstances a network may proceed through all its possible states before repeating any one of them. In practice, however, the network at some point hits a series of states around which it cycles repeatedly. Known as a state cycle, this repeated series of states is in effect an attractor in the system, like the whirlpool in the treacherous sea of complex systems dynamics. A network can be thought of as a complex dynamical system, and is likely to have many such attractors.

"I spent hours working on the networks by hand," Stu explained. "My pharmacology notebooks are full of them, all up and down the margins." Because the number of possible states even in small, modestly connected networks rises rapidly as you increase the number of elements, hand-calculated networks soon become unmanageable. To go beyond about eight elements, a computer would be necessary. "I got some guy to teach me to program, and prepared for my first run, a network with a hundred elements, each with two inputs, randomly assigned. That's why I had to shuffle the data cards." It was either a young man with extraordinary insight who entered the computer center that summer day of 1965; or a fool. Most experts would have said the latter, as even this modest network had some 10^{30} possible states, a mere hundred trillion times the age of the universe, measured at one state per second. The computer ran a good deal faster than one state a second. Even so, had the network ventured just the minutest way into its territory of total possible states before hitting a state cycle, the program would have run for days and Stu would have been bankrupt, as he was paying for the computer time himself.

"You had to have been extremely naïve to do what I did," Stu recounted with a wide grin. "But I was lucky. It went into a state cycle after going through just sixteen states, and the cycle itself was only four states. I said, 'Oh my God, I've found something profound.' I still think so. It's the crystallization of order out of massively disordered systems. It's order for free."

Stu was in the second year of his course at this point, medical studies barely on track while the obsession with Boolean networks deepened. When he wasn't computing networks he was exploring the literature, much of it foreign to him. Then, with a sense of profound shock, he came across a book published in 1963, called *Temporal Organization in Cells*. Its author, Brian Goodwin, had studied biology at McGill University, Canada, then mathematics at Oxford a few years earlier than Stu, and had pursued a doctorate at Edinburgh University under C. H. Waddington, one of the recent major figures of British biology. Waddington believed passionately that organisms must be studied as wholes, and that the principal challenge of biology was to understand the genesis of form. Entranced with this holistic approach, Brian integrated it with the molecular biology of Jacob and Monod, and produced a theory of how gene activity and oscillating levels of biochemicals could contribute to biological form. *Temporal Organization in Cells* was his thesis in book form.

"I thought, Oh shit, he's got there first," Stu recounted of his reaction when he first saw the book. "Then I thought, Hey, I don't understand this. What's it all about? Finally I said, 'He's got it wrong. Don't know why, but I'm sure he's got it wrong.' " The book was an attempt to show how molecular control systems, such as feedback, repression, control of enzyme activity—in other words, the intrinsic local logic of a complex system—gave rise naturally and spontaneously to oscillatory behavior and global patterns. Such behavior is an important component of living systems, such as circadian rhythms and the periodic activity of hormone and enzyme systems. The core of the book—the generation of order as an inevitable product of the dynamics of the system—resonated powerfully with Stu's view of the world. He immediately sent Brian a copy of the early results from the Boolean networks, but didn't enter into correspondence. "Stu's not the greatest correspondent, even worse than I am," Brian told me. "He prefers talking, and then it's pretty intense. A few hours with Stu is worth weeks with anyone else."

I knew what he meant. A conversation with Stu is like a match between a squirt gun and a fire hose: pretty much a one-way flow. But it is all worth it, every drop. By now, this conversation was in need of respite, and Stu and I took ourselves off to a small Indian restaurant near campus, the kind where you can write on the table. I asked speculatively whether the rise and fall of complex societies might also be described by the science of Complexity. Wild idea, I said, but I'd been reading studies of the inexorable rise and fall of civilizations through history, and the repeated pattern seemed to this naïve spectator to have the right "smell" about it. "Hadn't ever thought about it," Stu said after a few moments' reflection. "But, no, I don't think so." (Six months later he'd changed his mind, and called to tell me so triumphantly. At this lunch, however, he was much more focussed on the Boolean network story.)

I admitted that the order for free from the Boolean networks was impressive. But wasn't it little more than an analogy, a seductive image? "It's an analogy of sorts, of course, but it's deep. Listen to this." Stu went on to describe the countless experiments he'd run, which showed the emergence of two properties that were very biologic in character. The first concerned the number of cell types that are found in a range of different organisms of different complexity, and how they might be generated. The second related to the limited possibilities available to any cell type for changing into other cell types.

"I formed the strong conviction that the state cycles in my Boolean networks were equivalent to different cell types," he said. "That's obvious. That's just a restatement of my search for order. But then I started working out how many state cycles I got in networks with two connections. The number turned out to be very roughly the square root of the number of elements in the system. A network of 100 has about 8 state cycles, 8 attractors if you like; a 1,000-element network has about 33 attractors, and so on. A network with 100,000 elements, roughly the number of genes in the human genome, has about 370 attractors, and that's pretty close to the known number of cell types, 254." Early on in the work, Stu dug into the literature again, looking for information on the number of cell types in a range of organisms, in relation to the estimated number of genes. He got information for bacteria, yeast, algae, a fungus, jellyfish, annelid worms, and humans, representatives of different major groups, or phyla, that separated from each other in the Cambrian, 600 million years ago. The result was clear: the more genes an organism possessed, the more cell types it had. The fit was pretty good, not exact but within striking distance and therefore biologically reasonable. As significant, however, was the fact that the number of cell types in each organism was roughly the square root of the number of genes. Boolean networks and genomes both had the square-root rule.

"Now, either you've got to persuade me that 600 million years of evolution has independently honed genomes from different phyla so that they all generate cell types as the square root of the total number of genes," said Stu, "or you have to admit that my Boolean networks are more than just an interesting analogy, that there is something fundamental about the dynamics of this kind of system."

Stu asked me to imagine the networks as collections of light bulbs, red when they were fixed, blue when they were changing. In networks with two connections per element, large areas of light bulbs remain stable, unchanging for almost all attractors, with patches of change scattered among them. It was a graphic image, of twinkling blue islands in a sea of red. "There are two consequences of this pattern," explained Stu. "If this is a reasonable model for the generation of cell types, then all cell types should express most of the same genes, with only a small fraction different." That, as far as I knew, is correct.

"That's right. The second consequence—and this is the second of the two biological features of the networks—is that attractors resist perturbation: mutations in the islands don't propagate far. But when they do change, their options are limited to nearby attractors." I could imagine the twinkling islands behaving like this, their isolation in a frozen sea of red restricting what they might do in the face of perturbation. "Again, that's what you see in living systems. During development, cell types progress along highly constrained pathways. Once a cell has embarked upon a particular pathway it leaves behind many other options and the number of other cell types it can transform into is greatly diminished."

So, I asked, is the key property of the system that the local rules, the number of inputs each "gene" receives and the rules for responding to them, generates a global order in the system? "That's right." An emergent property? "Absolutely. It's unpredictable and counterintuitive. Order for free. Isn't it beautiful?" It was time to leave.

If random Boolean networks give insight into how cell types might be generated, what of the second major component of embryology, the way the cells assemble into a mature individual. "Go and see Brian Goodwin," Stu urged. "He's the poet of theoretical biology." The two men met first in 1967, at the Massachusetts Institute of Technology. Earlier that year, Stu had sent his Boolean network results to Warren McCulloch, a pioneer in network theory. McCulloch had written back promptly, saying with characteristic hyperbole, "All of Cambridge excited about your work," and inviting Stu and Elizabeth, his new bride, to spend three months in his large rambling house in Cambridge. MIT was the place to be, with all the great names in the study of parallel processing systems and some of the top people in theoretical biology, including Brian Goodwin for a while.

Stu and Brian immediately realized that their approaches to the search for order in biology were complementary, not conflicting, and they formed a strong personal and professional relationship. Brian was already becoming an important figure in theoretical biology, and soon emerged as a leading intellect. He's also known to occupy an extreme position in the intellectual spectrum.

"Poet of theoretical biology. Hmm. I guess Stu thinks I have a certain vision," offered Brian when we met in his office, now at the Open University in Milton Keynes, some fifty miles north of London. His tall bearing, distinguished graying, and Mediterranean good looks certainly fit the part of poet. "But, you know, a lot of people might think that's pejorative." I asked him about the dedication he'd penned in Stu's copy of *Temporal Organization in Cells*. It read: "There is no truth beyond magic." Meaning? "Two meanings really. One, when you've discovered the truth in science it does have the most extraordinary magical quality about it. It's the payoff, to recognize the deep order in biology, you feel you are in touch with something fundamental. But there's also a poetic sense in it: reality is strange. Many people think reality is prosaic. I don't. We don't explain things away in science. We get closer to the mystery." Sounds romantic, I ventured. "If it's a romantic view of science, so be it. It'd be a dull world without it."

It seemed to me that coupled oscillating chemical systems—the kind of mechanism Brian had explored in his book and that still interests him—were a long way from any notion of mystery. "Let me explain my approach, and then you'll see." The fundamental problem of biology is how you generate form, Brian began. "Long before Darwin, scholars were fascinated in biological form and how it related to the world. There were very different approaches to it: the Anglo-American approach, and the continental-European approach, particularly Germanic."

The first of these is rooted in the school of Natural Theology, which goes back to the eighteenth century. It focussed on function, how organisms worked. But its overarching ideology was that the wonders of the world provided evidence of a Divine Hand. "You'll remember the famous passage at the beginning of Paley's *Natural Theology*," Brian continued. "The story of someone finding a watch on the heath." I did indeed. I'd bought an 1854 edition many years ago—the first edition was published in 1802—and had read the opening passages several times, still discernible through foxing and erratic print. "In crossing a heath, suppose I pitch my foot against a stone, and were asked how this stone came to be there; I might possibly answer, that for anything I knew to the contrary, it had lain there forever," it begins. Then: "But suppose I had found a watch upon the ground, and it should be inquired how the watch happened to be in the place; I should hardly think of the answer which I had before given, that, for anything I knew, the watch might have always been there."

Paley goes on to explain, in a lengthy analogy, that the existence of the watch must imply an agent of design. "The inference, we think, is inevitable; that the watch must have had a maker; that there must have existed at some time, and at some place or other, an artificer or artificers, who formed it for the purpose we find it actually to answer; who comprehended its construction, and designed its use." The argument is then extended to biology: "The contrivances of nature surpass the contrivances of art, in complexity, subtilty, and curiosity of the mechanism . . . ; yet, in a multitude of cases, are not less evidently mechanical, not less evidently contrivances, not less evidently accommodated to their end, or suited to their office, than are the most perfect productions of human ingenuity." And what better example than the eye, asks Paley, so complex, so perfectly suited to its role?

"Ah yes, the eye," said Brian. "Even Darwin was worried about the eye." In *Origin of Species* he wrote: "To suppose that the eye, with all its inimitable contrivances for adjusting the focus to different distances, for admitting different amounts of light, and for the correction of spherical and chromatic aberration, could have been formed by natural selection, seems, I freely admit, absurd in the highest degree possible." Nevertheless, he concluded, that because there is "no limit to this power [of natural selection], in slowly and beautifully adapting each form to the most complex relations of life," the eye also is explicable by the slow, incremental assembly process that is natural selection.

So, while Paley explained the exquisite morphology of organisms in relation to their environment as evidence of Divine Design, Darwin explained it as the outcome of natural selection, the blind moment-to-moment adaptation of organisms to prevailing conditions, the products of random mutation sorted according to survival. "Both explanations focus on function, one is theological, the other scientific. But, I believe, the second is as wrong as the first—almost," said Brian. "And yet the notion of natural selection as an all-powerful force in generating biological order is now deeply rooted in our scientific culture. You only have to look at Richard Dawkins's much acclaimed *The Blind Watchmaker* to see that." The title is a nice turn on Paley's analogy of God the watchmaker, God the creator of nature. "This book is written in the conviction that our existence once presented the greatest of all mysteries, but that it is a mystery no longer because it is solved," begins *The Blind Watchmaker*. "Darwin and Wallace solved it, though we shall continue to add footnotes to their solutions for a while yet."

The book is an inventive and persuasive exposition of evolution by natural selection, and it leaves no significant room for other mechanisms. "It is the contention of the Darwinian world-view that . . . slow, gradual, cumulative natural selection is the ultimate explanation for our existence," writes Dawkins in his closing paragraph. "If there are versions of the evolution theory that deny slow gradualism, and deny the central role of natural selection, they may be true in particular cases. But they cannot be the whole truth, for they deny the very heart of the evolution theory, which gives it the power to dissolve astronomical improbabilities and explain prodigies of apparent miracle." Stuart Kauffman's random Boolean networks, it's worth noting, have the "power to dissolve astronomical improbabilities," in the absence of natural selection.

So, I asked Brian, are you looking at footnotes to Darwin's theory? "Indeed not. We need a new book."

Modern biology has all but lost any true notion of "the organism," lamented Brian. "The organism has been replaced by a collection of parts—genes, molecules, and the components that are supposed to make eyes, limbs, or whatever structure one is interested in." You mean the reductionist approach? "Exactly." But reductionism has been the triumph of modern biology, making biology more like physics, I suggested. Look at what we know about the structure of genes, how they are expressed, the incredible details of metabolic machinery that are now known. "All that is true. I don't deny those achievements. I just insist they tell you nothing important about biological form, how form is generated." He told me a favorite analogy: knowing the structure of H_20 gives you no clue as to why water goes down a plughole in a vortex. "We need a concept of the whole organism as the fundamental entity in biology and then understand how this generates parts that conform to its intrinsic order," he continued. That sounds a bit vague, I suggested. "Remember the Dahlem conference," Brian countered, referring to a scientific gathering we both attended in Berlin a decade ago.

The conference had been on evolution and development, and at one point there developed a vigorous battle over what you needed to know truly to understand how an organism is assembled. *Acrimonious* is a better adjective than *vigorous*. On one side were conventional molecular biologists, who insisted that when the DNA sequence of an organism was known, all would be evident. "The assembly instructions are written in the genes," was their position.

On the other side was Brian and a small group of unconventional molecular biologists, including Gunther Stent, an eminent researcher at the University of California, Berkeley. Stent was famous—or rather, infamous—among his colleagues for having suggested in a major review in *Science* that molecular biology had had its day, that it had no important intellectual challenges remaining. Other avenues of biological endeavor would tackle the real problems, like embryological development, he said. "Genes are only a start," was the Goodwin/Stent view; "without knowing the dynamics of the component parts you get nowhere." There was no resolution to the debate.

"Molecular biologists discovered that the linear sequence of nucleotides in DNA specifies precisely the linear sequence of amino acids in proteins," continued Brian. "No one doubts its importance. But they committed the error of imagining that, similarly, the linear sequence of genes in a genome specifies the genesis of form in an embryo, analogous with a computer program." He added emphatically: "There is no genetic program for development, no program that guides the system through its morphogenetic transitions." You're not saying genes are irrelevant? I asked, wondering how far this line of argument might go. "No I'm not. The genes set the parameter values." Meaning? "Meaning that they produce component parts of the system, within a range of values. The morphological transitions then are consequences of the cycle of dynamics generating geometry and geometry modifying dynamics. This gives us a 'free lunch' view of morphogenesis." Wait a minute, I pleaded. You're getting way ahead of me. Can we make this a bit more tangible, something I can understand intuitively?

"Sure I can. I'll show you the Acetabularia model."

Acetabularia acetabulum, a species of alga better known as mermaid's cap, lives in shallow waters around the shores of the Mediterranean. Its life cycle is regular and dramatic, beginning with the fusion of two "sex" cells to form a single cell, the zygote. This single cell develops rootlike structures, and then a growing stalk, which eventually reaches as much as five centimeters in length. The growing stalk produces rings of "hairs" near its tip, the so-called whorl. The whorl is subsequently shed as the tip opens up to form a disc, not unlike a mushroom cap.

"See this," Brian said as he indicated the whorl on a picture of *Acetabularia*. "It's a mystery, or at least it was. Doesn't have any known function, so why is it there? The Darwinian explanation—the functionalist explanation—would be that either we have failed to find its function, or that it once had a function but is now

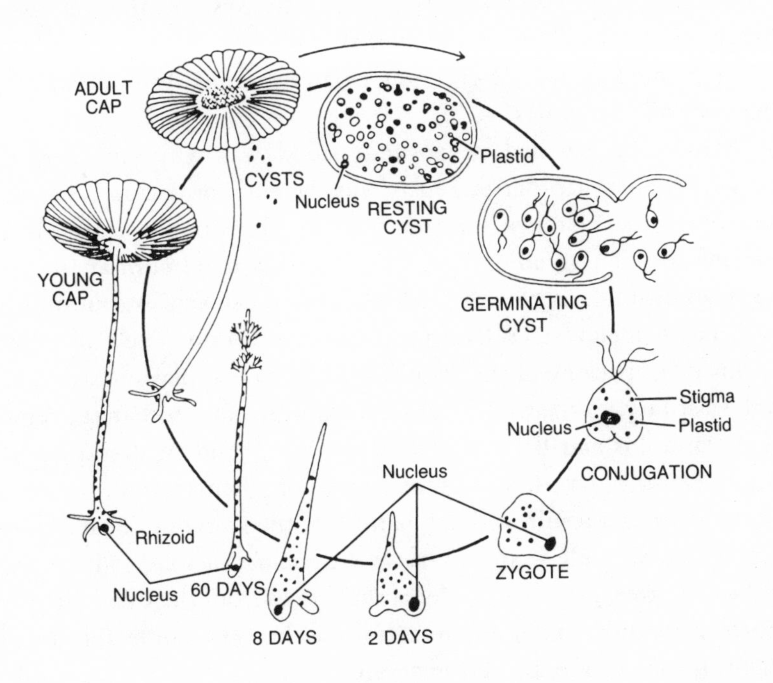

Fig. 2. The life cycle of *Acetabularia acetabulum*. Courtesy of Brian Goodwin.

vestigial. The truth is very different." Brian and several colleagues recently built a mathematical model of the development of the organism as a way of understanding the morphogenetic transitions, the major steps through which the developing organism passes. Essentially, the model includes aspects of calcium regulation in the cell, the changes of the mechanical state of the cytoplasm (elasticity and viscosity), and the response of the cell wall. These are the parameters of the system. "Biologically simple but mathematically complex," is how Brian described the model.

"Watch how the shape develops," Brian said as he set the model running on a video screen. "When we started work on this we expected that many patterns would be possible, that it would take us a long time to find the parameters that simulated *Acetabularia* morphogenesis." Not so. It all fell out very quickly, as if the *Acetabularia* form was a deep property of the system, like a ghost in the molecular machine. I watched as the schematic organism went through much of its life cycle, as Brian explained the unfolding of the system's dynamics.

"Something we had never understood was why and how the initially conical tip flattens just before whorl formation," he said, following the changes on the screen. "The model gave us an explanation. The gradient of calcium with a maximum at the pole becomes unstable as growth proceeds, and transforms into an annulus with a ring of elevated calcium and increased strain. The wall softens in this ring, because of the coupling between cytoplasmic strain and wall elasticity, so the wall develops maximum curvature in the region of the annulus and flattens toward the tip." And then you get the whorl forming, I said as I watched a ring of schematic hairs develop. "Yes." Just as a result of the dynamics of the system? "That's right. The reason why all members of the *Dasycladaceae* make whorls may be because they are a natural form that arises from the dynamic principles embodied in the organization of the cell. Some species use them, some do not, but all generate them."

I was struck by the parallel with Stu Kauffman's Boolean networks. Local rules generate global order? "Yes. It's an emergent property of dynamical systems. I view development as a dynamical system. Let's get back to the eye."

For various mechanical reasons, in *Acetabularia*, and in plants, generation of form is always accompanied by growth, a continual outward expansion. Because they are not so mechanically constrained, animal embryos can generate complexity in many more ways, including outward or inward deformation of sheets of cells, migration of cells, and other means. As a result, animals can produce tremendous internal complexity as well as intricate external pattern. In spite of these differences, the fundamental processes of developmental organization are the same as in the mermaid's cap, a dance of dynamical systems.

Briefly Brian explained the main events of morphogenesis in the animal embryo, which involves sequential processes of invagination and folding of sheets of cells that establish the ground plan of the organism's internal structure. "We can make computer models that simulate the process, using the same kinds of parameters as in the

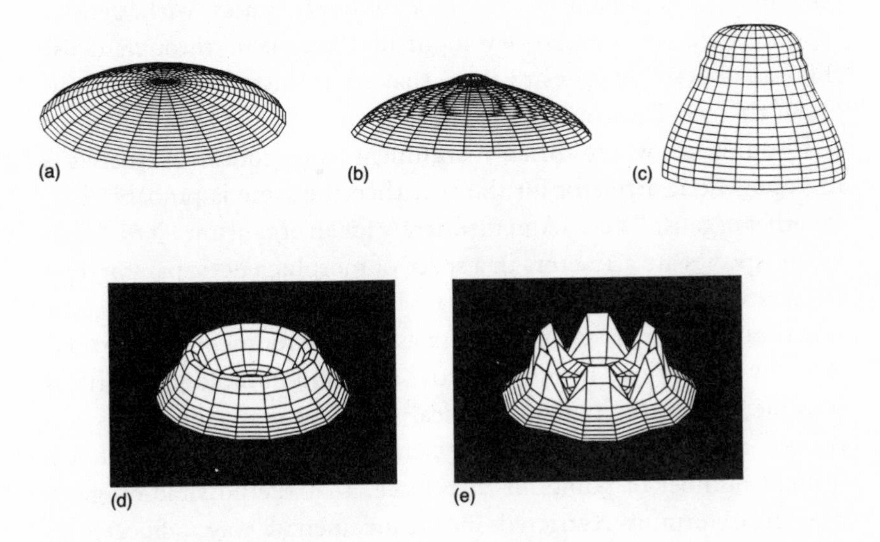

Fig. 3. Five stages in the development of the whorls in the computer-generated model of *Acetabularia*. Courtesy of Brian Goodwin.

Acetabularia model," said Brian, as we scanned pictures of these classic stages in development. He then described the developmental events that establish the form of the eye, processes of invagination and folding of sheets of cells. "Notice how the basic morphogenetic events for eye formation are simply repeats of the basic movements we've been talking about in other aspects of development?" Are you saying that, given the developmental processes that are known to operate, making an eye is easy? "Yes." Like making *Acetabularia* is easy? "That's what I'm saying, precisely. I'm saying that there's a large attractor in morphogenetic space that results in a functional visual system."

Eyes have evolved independently, oh, I don't know how many times, lots, because there's a morphogenetic attractor that specifies that kind of form? "More than forty times. Yes, *I* think so. Eyes are the product of high-probability spatial transformations of developing tissues. This is very different from the neo-Darwinian position, which states that organisms generate highly improbable structures, like the eye, that persist because they are useful. Natural selection holds organisms in these unlikely states with genetic programs that guide the developmental organism through dense thickets of possible states to those that are consistent with survival. Or so the neo-Darwinist's argument goes."

I could see where Brian's argument was going. If there's a morphogenetic attractor for the eye, then the same is probably true for other organs? "Yes." And ultimately for an organism? "Yes." You mean, species are attractors in a space of morphogenetic parameters? "That's what I mean. To a neo-Darwinist, every point in that space is realizable as an organism, as long as the environmental conditions favor its expression. In other words, any kind of biological form is possible, within certain mechanical limits. I'm saying that's not correct, that the organizational dynamics of morphogenesis define a limited number of points in that space, that the possible range of biological form is restricted in a fundamental way." Species as attractors in a dynamical system: it's a provocative notion, quite outside conventional biological thought. "You have to realize that what I'm saying is a plausible conjecture," conceded Brian. "But I think it's powerful, and will be shown to be more right than wrong."

Brian's position—and Stu Kauffman's, too—is the most recent expression of an intellectual tradition with roots in the eighteenth-century Enlightenment. Known generically as Rational Morphology—with a line of distinguished scholars from Kant and Goethe and running through Geoffroy Saint-Hilaire, Baron Georges Cuvier, William Bateson, Richard Owen, Hans Driesch, D'Arcy

Wentworth Thompson, Waddington, and others—the search was for "laws of form" that would explain the striking patterns of order seen in nature. Despite a diversity of approaches, the Rational Morphologists all had a deep conviction about the unity of the individual organism, and sought the generative source of order they discerned there. This is the second of the two great schools of interest in biological form, the first being the functionalists, with Natural Theology, Darwin, and Dawkins.

"Ours is a science of qualities, not quantities, and is therefore a Goethean science," said Brian, as we turned from the tangible to the philosophic. "Goethe is one of my heros in this respect." Responding to the suggestion that this may sound a little mystical to some ears, Brian said, "Maybe. But our approach views nature as intelligible. The creative principle of emergence is a deep mystery in many ways, it's true, and that's a property of complex dynamical systems. But ultimately it is intelligible. You can't say that about neo-Darwinism." François Jacob once likened natural selection to a tinkerer, a moment-to-moment tactician, cobbling together contraptions to cope with prevailing circumstance. It was meant as a description, not a criticism of the concept. "The problem of form is thus effectively 'reduced' to the problem of functional adaptation," said Brian. "It makes biological form unintelligible."

I wondered whether there was any place for natural selection in Brian's view of the world. "I'm not denying natural selection," he said. "I'm saying that it does not explain the origins of biological form, of the pervasive order we see out there." On a scale of one to ten, he said he would rate the importance of natural selection—in the context of the generation of form—as close to one. "And Stu's position is the same, you know. But I don't think he's ready to say all this transforms neo-Darwinism. Logically, what he's saying does, but he allows natural selection a greater role than I do."

"When I first wrote up my Boolean network results for a scientific paper, I pretty much dismissed natural selection as having any importance at all," Stu told me. "Just look at the quote at the beginning of the paper." It read: "The world is either the effect of cause or chance. If the latter, it is a world for all that, that is to say, it is a regular and beautiful structure." *Marcus Aurelius*. Marcus Aurelius was an Emperor of Rome. "I liked the quote because, there I was, assembling networks at random, and despite that we see all this order," he explained. "You know that phrase of Einstein's, about 'searching for the secrets of the Old One.' Well, I thought that the Old One wouldn't fool around, that there'd be some deep logic out there, and I thought I'd glimpsed it in the random Boolean nets. And Marcus Aurelius seemed to be saying that it's OK if it's random. It's still beautiful. And, no, I didn't think I had to worry about natural selection, but I do now."

John Maynard Smith was responsible for twisting Stu's arm on this. John is Britain's most eminent evolutionary biologist, a champion of neo-Darwinism, with a strong mathematical bent. With the appearance of the classic absent-minded professor, equipped with penny glasses and long white hair, John has the none-too-common combination of the keenest of critical intellects and great professional generosity. John and Brian were close colleagues in the school of biological sciences at the University of Sussex for many years, and two more different views of the world of nature could hardly be imagined. John is also an enthusiastic gardener, and opens his garden to public view. It's an easy and very pleasing excursion to visit John's garden in Sussex in the morning and Darwin's in nearby Kent in the afternoon, thus satisfying scientific and horticultural interests simultaneously.

Impressed as he was by the emergence of order in Stu Kauffman's random Boolean networks, John nevertheless saw them as incomplete. "Until you put selection into these models they have absolutely nothing to do with life," John told me when I visited him at the University of Sussex. "They're just not interesting to a biologist. Stu's learnt that now." The learning process began when the two men first met, in 1968 at a theoretical biology conference, held at the Rockefeller Foundation's Villa Serbelloni, on the shore of Lake Como, Italy. Then, and each time they met as the years passed, John would try to persuade Stu of the importance of natural selection in shaping biological systems. On one occasion, ten years ago, the two were walking on the South Downs, close to the University of Sussex, when John said: "By and large, those who held that selection played a major role in evolution were English country gentlemen, but, forgive me Stuart, those who have not have largely been urban Jews."

I asked John what he meant. "English country gentlemen is too narrow; European, perhaps, and ladies, too," he began. "People like Darwin and Wallace, they were country boys, and developed a passion for natural history. But because they were intellectuals they became interested in how it came about, the incredible functional adaptation you see. You can't study nature without knowing there are bizarre adaptations out there, complicated ways of life that seem to fit an organism to its environment. So the problem becomes, how do I explain it? Adaptation by natural selection is the answer." I could appreciate the argument about English country gentlemen, I said, but what about urban Jews? "I mean urban intellectuals, people like Stu Kauffman and Steve Gould. It's the search for universal truths. They seem to say, if there are no universal truths, how can you do science? Natural selection appears to be too ad hoc for them, just opportunistic adaptation. For me, that's the way nature is."

I asked Stu if he is indeed looking for universal truths. "What I'm looking for is a deep theory of order in biology. If you view the world as John does, then our only option as biologists is the systematic analysis of ultimately accidental machines and their ultimately accidental evolutionary histories. I know there is more to it than that." And natural selection: how good a job did John do in arguing its importance? "The theory of natural selection is brilliant, no question. And, I know Brian doesn't agree with me, but it is an important force in evolution: say five on your scale of one to ten. But there are things Darwin couldn't have known. One of them was self-organization in complex dynamical systems. If the new science of Complexity succeeds, it will broker a marriage between self-organization and selection. It'll be a physics of biology."

Biologists will find it tough enough to assimilate the notion of self-organization into their current worldview. "There's more to come," said Stu. "There's the edge of chaos."

CHAPTER THREE

Edge of Chaos Discovered

"The edge of chaos." Now that's an intriguing phrase. "It's more than intriguing," said Stu Kauffman as we settled back in his office, there to continue a conversation that was constantly interrupted by telephone calls finalizing what sounded like complex real-estate deals. Stu likes to juggle many different things at once. "It's a beautiful phrase, and it may just be fundamental to this science of Complexity." And, he added, it may be fundamental to the world out there, waving an arm in the direction of a window that the University of Pennsylvania had apparently not washed in many a year. By "out there" Stu meant nature. *All* of nature.

How, I asked, do we get from random Boolean networks to this mysterious-sounding territory, the edge of chaos? "A long story," Stu replied. "Remember I told you about the three months I spent at MIT, with Warren McCulloch? It was a tremendous time, so intense, so exciting. I was just a medical student, and all these people were smart as hell, and famous." Stu was smart, too, McCulloch recognized that, and so guided the young Kauffman through what was, to him, virgin mathematical territory. "Warren was what you need in a mentor: enthusiastic about my science, willing to help when it was needed, willing to recede when it was time for credit to be distributed, willing to share authorship on a paper if he thought it would protect me." Pretty unusual in my experience, I observed.

McCulloch did coauthor a paper with Stu, technically the first report of the Boolean network results; but it was an internal MIT document, not a traditional scientific paper. "Warren," Stu asked his mentor as they prepared the report, "is anybody going to care about all this?" "Yes, but it will be twenty years before anyone will take any notice," McCulloch replied without hesitation. "I couldn't believe it," Stu told me, as he recalled the shock of that moment, some two decades earlier. "Twenty years sounded like forever to me. I knew this thing I'd found was profound, had profound implications for the way organisms were made, and that it would shake up biology. Surely, I thought, people would jump up and down and shout, 'Hallelujah.' You have to remember, I was only twenty-eight at the time. And naïve."

After collecting a medical degree from San Francisco in 1968, finishing in the lower echelons of his class, Stu did a year's internship at Cincinnati General Hospital, then became a faculty member in theoretical biology at the University of Chicago for four years, on to the National Institutes of Health, in Bethesda, arriving finally at the University of Pennsylvania in 1975, where he stayed for sixteen years before installing himself more or less full-time at the Santa Fe Institute in 1991. The odyssey had taken him from theoretical biology to bench biology, from abstract random networks to the genetics of fruit flies, *Drosophila*. "I became a real biologist," he told me with considerable mirth. (Stu, I knew, has no great reputation as an experimental scientist.)

During a brief visit to the University of Chicago in 1970, John Maynard Smith offered to teach Stu "how to do some sums." John was fascinated by the Boolean networks, in spite of their—to him—distance from biological reality. So, with John's considerable mathematical expertise as encouragement, Stu analyzed the networks more deeply. He already knew that when each element in the network had only one connection from other elements, nothing interesting happened. The system virtually froze. He also knew that with four or more connections the system became unstable, chaotic. Nothing of interest here. And of course he knew that with just two connections, something very significant happened: a small number of attractors he took to be analogous to cell types were generated. But he did not realize just how significant this intermediate territory—between one and four connections—really was. Stu had traversed the edge of chaos, but had been unaware of it. As a result, interest in random Boolean networks languished for almost two decades.

While Stu was playing the part of real biologist, the worlds of computing, mathematics, and physics were steadily turning their collective attention to dynamical systems. Arcane new worlds of neural networks, spin glasses, genetic algorithms, and chaos theory opened up intellectual horizons, offering fleeting glimpses of complexity and ways of understanding it. The development of novel mathematical techniques and the necessary computing power to cope technically with the demands of dynamical systems was central to this dispersed effort. Equally important, however, was the nascence of an outlook that, fundamentally, these many activities draw upon common concepts. This new intellectual climate nurtured a slow rekindling of interest in Boolean networks, and a rebirth of fascination with a phenomenon known as cellular automata. Both would eventually lead to the discovery of the edge of chaos.

John von Neumann, the brilliant Hungarian mathematician, invented cellular automata in the 1950s, during his quest for self-reproducing machines. Cellular automata, the computer jock's equivalent of a menagerie, are a kind of complex dynamical system. Imagine an infinitely large grid of squares, like endless graph paper. Each of the squares, or cells, may be either black or white, depending on the activity of neighboring cells. Simple rules govern the state of each cell, such as, if four or more of a cell's contiguous eight cells are white, then the central cell changes state. Like Boolean networks, cellular automata progress through a series of states, at which each cell examines the activity of its neighbors, and reacts according to its rules. Complex, dynamic patterns develop and roam across the entire grid, the nature of which is influenced but not tightly determined in detail, according to the activity rules.

Notice that global structure emerges from local activity rules, a characteristic of complex systems.

When he first came across cellular automata and their potential for self-reproduction, Chris Langton was hooked. This was while he was at the University of Arizona, Tucson, in 1979. Chris had recently become obsessed with creating artificial life, and the computer was going to be its medium.

Brought up in Lincoln, Massachusetts, the oldest of three sons of scientist parents, Chris was an archetypal rebel of the sixties. Long haired, bejeaned and guitar playing, he had about as fragmented a university career as could be imagined: he attended Rockford College, Boston University, the University of Arizona, and the University of Michigan, each of which was interrupted variously by anti–Vietnam War protests; work at Massachusetts General Hospital as quid pro quo as a conscientious objector; a spell helping out at a primate research colony in Puerto Rico; a rewarding job as a carpenter with a building contractor; and a brief partnership in a stained-glass workshop. And a lot of time on the road.

In 1975 a near fatal hang-gliding accident on Grandfather Mountain, North Carolina, shattered dozens of bones, including both legs, all but detached his right arm, punctured a lung, crushed his face on impact with his knee, and inflicted generalized brain trauma. When eventually he recovered he went to Tucson, in the fall of 1976, where he planned to study astronomy but soon decided instead he wanted to combine courses in evolutionary biology, computing, and anthropology. Without articulating it, Chris, like Stu Kauffman, also was seeking the secrets of the Old One. "There I was, this misshapen body, skeletal almost, raving about these ideas I had, that I could model cultural evolution on a computer," he told me when we met recently at the Santa Fe Institute. "Boy, did I look the part of the unhinged lunatic."

Even now, fifteen years on, ideas flood faster than words can cope, and so a conversation with Chris inevitably constitutes excursions first in one direction, suddenly to be interrupted by one side issue, then another, a brief revisit to the main theme, and more diversions. Given sufficient time, tremendous territory is covered and the listener is left in no doubt that here is a mind that can't help looking for, and finding, connections. I first met Chris at the Southwest Prehistory conference at the institute, and he explained that, after Tucson, he had gone to the University of Michigan, in 1982, officially to pursue a doctorate in the dynamics of cellular automata, but still driven by the notion of artificial life. He still wears jeans, still sports long hair, and still plays guitar, but he's added the silver and turquoise accoutrements of the Southwest: buckle, bracelet, and bolo tie. His face has the rugged look of the outdoors, with no clue to the horrendous trauma it suffered.

Before leaving Tucson, Chris had already been following in von Neumann's path with cellular automata as much as he was able, following other paths that looked similar, always seeking the goal of self-reproduction and complexity in the computer. What had attracted him to Michigan was the statement of purpose of the computer science department: "The proper domain of computer science is information processing writ large across all of nature." All of nature? "Yes, anywhere information is processed," said Chris. "Information's the key."

Enjoying the technical resources of the Michigan environment—an Apollo workstation rather than the Apple he'd been using—and the von Neumann intellectual legacy, Chris immersed himself in the dynamics of cellular automata. "That's how I got to the edge of chaos stuff . . . met Steve Wolfram . . . heard about his four classes . . . the meaning of universal computation . . . nonlinear dynamics, chaos . . . Wolfram didn't establish a relationship between the classes . . ." Whoa, I pleaded. Who's Wolfram, and what are these four classes? "Wolfram's really bright, and was partly responsible for the reemergence of interest in cellular automata," Chris explained, slowing down. "He's an entrepreneur, and Cal Tech didn't like that—he was junior faculty at the time—so he left, and went to Princeton, the Institute for Advanced Study. That's where he came up with this classification of C.A. behavior."

Mathematicians already knew that many dynamical systems exhibit three classes of behavior: fixed point, periodic, and chaotic. When Wolfram tested cellular automata behavior, he found these same three classes, which he labelled one, two, and three. But he also came across a fourth type—class four—which was intermediate between chaotic and fixed or periodic behavior. "Class four behavior is the most interesting," said Chris. "You can get universal computing there."

To someone like me, to whom computing is what happens when you tell a computer to do something specific, the notion of a protean pattern on a computer screen doing computing is a challenge. "Think of it as manipulating information, complex manipulation," Chris tried helpfully. He told me about the Game of Life, effectively a set of cellular automata rules that generates endless, bizarre, and often lifelike patterns. Invented by British mathematician John Conway in the late 1960s, Life, as people call it, conformed to a theoretical prediction made by fellow Brit Alan Turing twenty years earlier. He had invented the principle of universal computation, and eventually demonstrated it on a simple device known as a Turing Machine. Embodying the principle of all possible computers—hence the term universal computation—the Turing Machine was able to manipulate information in complex ways. And so, too, was Life, all by itself. It is truly gripping to behold as it unfolds on a computer screen, and I know of no one in or outside "the business" who can watch it with indifference.

If Steve Wolfram had identified the four classes of cellular automata behavior, I asked Chris, how did you manage to take it further? "Steve was working with a fairly limited system and was only able to sample discrete behaviors," he explained. "Think of it like this. He was using a probe, sticking it in here, then here, then here, and analyzing the behavior at each point." Characteristically, Chris was at the board busily sketching, giving me verbal and visual descriptions simultaneously. "I knew that with dynamical systems you can sometimes identify a parameter to make the system exhibit the whole spectrum of behaviors, exploring them all. I wanted to do the same thing for cellular automata, to move smoothly through the space of rules, watching the change in behavior as I went." A broad stroke passed through the four classes, one, two, four, three. "Like that," said Chris, describing what he eventually found.

Chris developed something he called the lambda parameter to do the job. Difficult to cast in any tangible analogy, lambda is a mathematical device that sets the rules of the cellular automaton and allows the consequences to be monitored across a continuum. Like a demon in the machine, I ventured? "You could say that." Chris prefers precise mathematical notation, and is constantly surprised when his audience has no idea what he's talking about. ("A power law distribution? In English? Oh, it's the probability that something is going to happen is one over some number to some power. The question is, what is the power?" No, no, Chris. In *English*. . . .) This time, however, he's simply recalling one of those strokes of luck out of which deep discoveries sometimes flow. "I set lambda at 50 percent, generated some rule tables, and expected them to come out in the chaotic region. But every one of them looked just like the Game of Life, all this interesting behavior. I said, 'That can't be true; something's wrong with my system.' Turns out that by mistake I'd set lambda at 30 percent, not 50." By accident, Chris had homed straight in on Wolfram's class four region, the region where maximum computational capability resides. But, I asked, hadn't Wolfram already demonstrated that? "Steve knew that kind of behavior existed, but he didn't have an overall structure of the behaviors in rule space," said Chris. "I was able to roam around in the space of rules and find out where the different classes of behavior were, where this particularly interesting class of behavior, class four, was." Chris had effectively produced a topography of cellular automata behavior, and dynamical systems behavior in general. For all anyone could have predicted, the different kinds of behavior—frozen, chaotic, and intermediate—could have been scattered haphazardly in the system. But Chris had visited that space of rules, explored it, and saw that as you leave ordered territory and enter the region of chaos you traverse maximum computational capacity, maximum information manipulation. "I discovered it was a very narrow region located between class two and class three behavior," said Chris.

In this most intangible of worlds, the rule space of cellular automata, we were talking about moving from one region to another, crossing a no-man's-land, where chaos and stability pull in opposite directions. It seemed like an Alice in Wonderland world, unreal, bizarre, a place where strange things happen. Chris referred to this no-man's-land as "the onset of chaos." But, he realized, this wasn't just an unreal, Alice in Wonderland world. It was the real world. He began to see the switch from order to chaotic regimes in dynamical systems as analogous with phase transitions in physical systems, the switch from one state to another; from the solid state to the gaseous state, for instance, perhaps with a fluid intermediate state. The notion of analogous phenomenology between universal computing and these physical phase transitions wormed its way into Chris's mind. Here was a shift from abstract computation to the reality of the physical world.

"You see phase transitions all the time in the physical world," said Chris. "Did you know that cell membranes are barely poised between a solid and liquid state?" I did, but hadn't thought of it in these dynamical terms. "Twitch it ever so slightly, change the cholesterol composition a bit, change the fatty acid composition just a bit, let a single protein molecule bind with a receptor on the membrane, and you can produce big changes, biologically useful changes." I asked whether he was saying the biological membranes are at the edge of chaos, and that's no accident. "I am. I'm saying that the edge of chaos is where information gets its foot in the door in the physical world, where it gets the upper hand over energy. Being at the transition point between order and chaos not only buys you exquisite control—small input/big change—but it also buys you the possibility that information processing can become an important part of the dynamics of the system."

Chris had made his serendipitous discovery soon after he signed on at Michigan. He then spent two years exploring all kinds of parameter settings, to get a feel for the space, to get a feel for the power that emerges where order and chaos meet. He was unaware that Norman Packard was close on his heels, a second adventurer in a strange land.

* * *

Norman Packard, a native of Montana, had been part of an adventurous group of physicists and mathematicians at the University of California, Santa Cruz, who, in the 1970s, effectively solved the puzzle of chaos. Known as the Dynamical Systems Collective, the group was considered by many older and supposedly wiser folk to be wasting its time and talents worrying about chaos, which, "everyone knew," was mathematically intractable and uninteresting. Norman, his close friend Doyne Farmer, and others in the group also devoted hours to developing computer-assisted methods of winning at the roulette table, thus confirming more conventional minds in their opinion that the people of the Dynamical Systems Collective were at best misguided in their academic interests. The collective eventually won recognition and respect as chaos theory emerged, and its members dispersed to respectable centers of research, Doyne to Los Alamos National Laboratory and Norman to the Institute for Advanced Study at Princeton.

Steve Wolfram had invited Norman to join him at Princeton, which he did with enthusiasm. "I was interested in evolutionary dynamics and the creative aspects of chaos," Norman told me. "There's an analogy between them. Chaos creates this infinity of patterns, and you never know what will happen next. And there's the creativity of evolution, starting with a chemical soup billions of years ago, and here we are now, thinking about it all." We were talking in the Santa Fe office of the Prediction Company, a newly formed corporation that aims to bring the power of dynamical systems research to analyzing and predicting the movement of financial markets, stocks, bonds, and currency. Norman and Doyne Farmer are joint scientific chiefs of the venture. The office is a modest wood frame house near the city's historic plaza, with white walls, wooden floors, piles of the *Wall Street Journal* lying on a low table, and a framed picture of Einstein leaning by a door. "Cellular automata have a rich array of possible dynamical behaviors, and so I saw this as a way of exploring my interests," explained Norman, after a phone call brought the news of new financial backing for the company's technical adventure.

Norman went to Princeton in 1983, a year after Chris Langton started his thesis work at Michigan. Like Chris, Norman also began to study cellular automata rules, "roaming around in this space of possible rules." Using a different approach from Chris's lambda parameter, Norman also explored the topography of cellular automata behavior. He, too, discovered the narrow transition region between order and chaos, and realized its potential for complex information manipulation. Two researchers, exploring similar territory, unaware of each other, but reaching the same destination.

"We met at the Evolution, Games, and Learning conference," said Norman. "I didn't fully understand Chris's lambda parameter at first, but it seemed to me that we were both looking at the same phenomenon." Norman had been co-organizer of the conference with Doyne Farmer, held in May 1985, at the Los Alamos National Laboratory. The two wrote an introduction to the subsequent conference volume, which described the possible commonalities between evolutionary dynamics and dynamical systems, particularly complex adaptive systems. Norman was also coauthor of two other papers. But he didn't write up his talk about the cellular automata work. "Too many other things going on," he explained. "I'm always pissed at myself when I do research and then don't write it up. Happens too often, I'm afraid." Chris did write a paper, called "Studying Artificial Life with Cellular Automata," which reflected his continued interest in artificial life, but also described his lambda parameter and the discovery of the onset of chaos.

With nothing in the written record about Norman's presentation, there is no objective way to compare the state of development of Chris's and Norman's research at that time. The way Chris remembers it, however, is that he was ahead, that Norman had not yet made the vital connection between class four behavior and the transition between order and chaos. "I remember driving back from the meeting, thinking to myself, 'Boy, I really told those guys something.' " In any case, both men were on the same intellectual track and arrived at the same place. Whereas Chris had termed the transition point "the onset of chaos," Norman coined the phrase "the edge of chaos." It is much more evocative, and brings forth images of being poised in space, tentative, dangerous even, yet full of potential. Like all powerful phrases, the edge of chaos has stuck, and has become iconic for the immanent creativity of complex systems.

The discovery that universal computation is poised between order and chaos in dynamical systems was important in itself, with its analogies to phase transitions in the physical world. It would be interesting enough if adaptive complex systems inescapably were located at the edge of chaos, the place of maximum capacity for information computation. The world could then be seen to be exploiting the creative dynamics of complex systems, but with no choice in the matter. But what if such systems actually got themselves to the edge of chaos, moved in parameter space to the place of maximum information processing? That would be really interesting: the ghost in the machine would seem to be almost purposeful, piloting the system to maximum creativity.

"It was my intuition, and it was Chris's, too, that the edge of chaos could be useful for evolutionary purposes," explained Norman. "I wanted to show that this was true, that systems adapt toward the edge of chaos. The logic was that if computations are seen to be good in an evolutionary context, you should get yourself to dynamical interactions at the edge of chaos." He decided to play God, albeit one with modest goals.

He established a set of rules for a cellular automaton, allowed them to mutate using a genetic algorithm, and set them the task of a particular computation. I asked whether he'd expect the rules to improve through natural selection. "That was the idea," said Norman. So you'd assign these "improved" rules a higher fitness in the game? And you'd expect the fittest rules to be generated at the edge of chaos? "Yes. I felt I was controlling things more than I would have liked, but I knew that it might show what I was looking for." It did. "The population of rules is seen to move toward a region in the space of all rules that marks the boundary between chaotic rules and non-chaotic rules," Norman wrote in the scientific paper that described the work, published in 1988. The paper was titled "Adaptation Toward the Edge of Chaos." It is a landmark in the emerging science of complex adaptive systems.

The discovery of the edge of chaos in the behavior of cellular automata was a vital step in this process, but in a sense was simply an echo of what mathematicians already knew about dynamical systems in general. The notion of universal computation at the edge, however, was definitely a new wrinkle. And, conceptually, making an explicit analogy between the dynamics of the edge of chaos and phase transitions in the physical world, as Chris had done, was a breakthrough. But without doubt the crowning achievement was the demonstration that a complex adaptive system (Norman's cellular automata with the assigned computation task) not only moved toward the edge of chaos but also honed the efficiency of its rules as it went.

Shortly before Norman's paper came out he visited Stu Kauffman in Santa Fe. "We were sitting in Stu's hot tub one evening, and I was telling him about these results, computation at the edge of chaos," explained Norman. "Stu got very excited and shouted, 'That makes perfect sense in the context of my Boolean networks. . . . It's all the same thing, the same goddamn thing.' "

In 1985 Stu had spent a sabbatical leave in Geneva, during which time he visited Paris, where he met Gerard Weishbuch, a physicist at the École Normale Supérior. Stu had recently started to tinker with Boolean networks again, wondering whether he might be able to see adaptation. It turned out that Boolean networks were experiencing a renaissance, at least in Europe. Weishbuch shared an office with Bernard Derrida, another physicist, who worked on what he termed "Kauffman networks." That must have been a surprise, I said. "It was. All this work going on, just burgeoning. I was delighted."

Back in the San Francisco days, Stu had "tuned" his network by changing the number of connections each element received, one, two, three, four, sometimes with as many connections as there were elements. Pretty crude, but effective for what he was doing at the time. He had seen states of pervasive order, with one connection, when most of the "light bulbs" were frozen as an island of red. He'd seen chaos, with many connections, when kaleidoscopic patterns of red and blue surged wildly through the system. And he'd seen that with two connections interesting structure emerged, twinkling blue islands in a red sea. But, like Wolfram and the four classes of cellular automata rules, Stu had not got a sense of the overall topography and its significance. Paris would change that. "Bernard was tuning a different parameter from mine, more subtle," Stu explained. "He'd begun to see the same phenomenon that Chris Langton had seen with cellular automata, the edge of chaos. Blue islands, shimmering, changing, in tenuous contact with each other."

Later that year Stu worked with Weishbuch, refining Derrida's approach, and began to develop a sense of the topography of the different states of the system, of complex computational capacity lying between ordered and chaotic regimes. "Gerard and I didn't write up the work," lamented Stu. "Nobody would have cared." He cares, however. And well he might, as the edge of chaos notion is likely to be extremely important in the world of complex adaptive systems. Priority of discovery is, at the very least, a matter of professional pride, and more likely an accomplishment deserving of serious recognition. The only explicit reference in the literature to Stu's claim to have independently discovered the edge of chaos is in his own book, *The Origins of Order*, published in the summer of 1992. In a late draft of the book, a discussion of the importance of the edge of chaos stated: "This suggestion has been made by myself in 1985, by C. Langton (1990), N. Packard (1988), and most recently by J. Crutchfield (personal communication 1990)." This 1985 citation to himself refers to the discussions Stu now describes as having taken place in Paris that year, not to a publication.

By citing a 1990 paper of Chris Langton's in this passage rather than the 1986 paper, Stu appeared to be giving priority of discovery to Norman Packard. Norman concedes that Chris was first. In any case, neither Chris nor Norman can recall Stu referring explicitly to the edge of chaos phenomenon until after they independently had done so and had talked to Stu about it. "I simply forgot that the edge of chaos was an interesting place to be," Stu told me.

* * *

When he returned to Philadelphia after his European sojourn, Stu turned at least part of his attention to evolution, specifically to adaptation. Four years of hectic research followed, which brought the edge of chaos notion to the verge of real biology. It began with fitness landscapes, a concept that the University of Chicago geneticist Sewell Wright developed in the 1930s. The imagery is deceptively simple.

You have to think about the "fitness" of an individual in terms of different combinations of gene variants it might have. Now think of a landscape, in which each different point on the landscape represents slightly different packages of these variants. Lastly, if you imagine some of the packages as being fitter than others, raise them up as peaks. The fittest of the packages has the highest peak. The landscape overall will be rugged, with peaks of different height, separated by valleys. Remember, this landscape represents fitness probabilities, places where individuals of a species might be, depending on the combination of genetic variants they have in their chromosomes. If an individual happens to be in a fitness valley, then mutation and selection might push it up a local peak, representing a rise in fitness. Once on the local peak it may, metaphorically, gaze enviously at a nearby peak, but be unable to reach it because that would require crossing a valley of lower fitness.

"It's a pretty image," said Stu. "I love it." Stu developed the idea further, and imposed upon it the structure of Boolean networks: fitness was determined by the number of genes in the species (the elements in the network) and their interactions (the number of connections between the elements). By tuning the connectedness of the genes, fitness of various combinations changed, thus changing the topography of the landscape. Working with Simon Levin, a biologist at Cornell University, Stu used the tunable fitness landscape concept to show that, powerful though natural selection may be in some cases, it is often unable to move a species toward fitness peaks and that the dynamics of the genetic system itself may exert a strong influence in this respect. "So, I guess I'm grateful to John [Maynard Smith] for getting me to think about selection," said Stu, "but I'm pleased to see that it has its limits, just as I always suspected."

The fitness landscape notion moved further toward imbuing the edge of chaos with biological reality when Stu linked two landscapes together. "Imagine a fly," said Stu. "It has a fitness landscape. Now imagine a frog. It has a fitness landscape, too. But they're not independent. The frog shoots out its tongue, zap, the fly's gone. That's part of life. Now suppose the fly evolves slippery feet so that the frog's tongue doesn't stick. The frog goes without dinner, and its peak on the fitness landscape goes down: it's less fit. The fly is fitter, and so its peak rises. So the coupled landscapes change, each responding to the other." The next step in the story is that the frog evolves hairs on its tongue—or some such device—and is able to catch the fly again. Fitnesses change, landscapes change. "It's the classic biological arms race," explained Stu. "Predator and prey constantly trying to be one step ahead of the other."

Biologists call the phenomenon the Red Queen effect, so named by Leigh Van Valen of the University of Chicago as resonant with Alice in *Through the Looking-Glass*: the predator and prey species have to keep running hard (evolutionarily) just to stay in the same place. It's an apt analogy, as the species inhabit fitness landscapes whose topographies are constantly changing, very much what you would expect in a Looking Glass world. The Red Queen effect is particularly pertinent in biology, because it is a reminder that species do not lead isolated lives but instead are linked inextricably with others. The evolutionary success of one species may therefore be as much a function of what other species do as what the species itself does. Some biologists go so far as to argue that the Red Queen is *the* driving force in evolutionary history, with environmental change playing only a minor part. The notion is clearly resonant with the dynamics of complex systems, an internal rather than an external engine for change for the species as a community.

"Now imagine that instead of two species you've got a hundred," said Stu, warming to the potential complexity of such a system. "That's a hundred coupled landscapes, interactions all over the place." I tried to imagine it, but the simple, vivid image of the fly and the frog evaporated and was replaced by confusion. Anything could happen, I said. "Anything could, but it doesn't," replied Stu. "We tune the interactions—internal, between the genes in the species, and external, how one species impinges on another—we watch how the system works, how the average fitness changes with different combinations of interaction. Guess what happens." I didn't have to. "The system moves through activity states, maybe frozen, may be chaotic, but eventually it comes to rest, with fitness optimized, poised at the edge of chaos."

That, I said skeptically, sounds like group selection. "It *sounds* like it, but it's not," Stu shot back. It was once thought that individuals within a species, or species in a group, might shape their behavior for the good of the group. Nowadays, biologists realize that individuals act in narrow, Darwinian, selfish ways, and will cheat if they can. The suggestion that a group of species might adapt collectively, with group benefit a goal, causes pitying smiles to come to biologists' faces. "But you see, the individual species in my group *are* behaving selfishly," said Stu. "That's the beauty of it. Collective adaptation to selfish ends produces the maximum average fitness, each species in the context of others. As if by an invisible hand—Adam Smith's phrase about markets in a capitalist economy—collective good is ensured."

It looked almost too good to be true. I tried again to imagine an ecosystem with many species interacting, each pursuing its own evolutionary ends, each evolutionarily tuning its own genetic connections and its interactions with other species, the result of which is that the community settles to a position of maximum sustained fitness. Some of the species would be hovering in a kind of evolutionary equilibrium while others among them engage in Red Queen antics; but all are components of a system delicately poised. Suddenly I saw that by tuning their interactions, species effectively were honing their ability to evolve. That would be astonishing. Are you telling me that your creatures get better at evolving in the midst of all this activity, that they improve their evolvability? "Yes," he said with a wide grin. "Isn't it gorgeous?"

It certainly looked gorgeous. Then Stu stunned me with something completely unexpected. "You know what Phil Anderson said when I talked about this at the institute?" Stu asked. "He said, 'That's mini-Gaia.' " It's what? "Mini-Gaia." Are you serious? I asked. Philip Anderson is a Nobel Prize–winning physicist at Princeton, with close links with the institute. He's nobody's fool. And the notion of Gaia, Earth Goddess, as a superorganism maintaining global balance, to many scientists is less than respectable. "Sure I'm serious. Phil said mini-Gaia, and I think he's right." In my exploration of complexity and how it might illuminate some of the greater patterns in nature, I'd developed some expectations of where it might take me. Embryological development, evolution, ecosystems, social complexity—but never once had Gaia crossed my mind. And yet it immediately made sense: here, in Stu's complex computer model, an ecosystem brought itself to a collectively beneficial state, control though vast networks of interactions. Certainly sounded Gaia-like. I made a mental note to pursue Gaia later.

I asked how he could be sure his computer ecosystem comes to the edge of chaos. "We follow the dynamics, see the system when it's frozen, when its chaotic, and we can see that it settles down in this intermediate state, with high fitness," Stu explained patiently, repeating what he'd said earlier. He then conjured up the light bulb analogy again, finishing with the ecosystem as represented by shimmering, barely changing blue islands, tenuously touching each other. Its familiarity was reassuring. "What's wonderful is that you can actually see adaptation getting the system to the edge of chaos," continued Stu. "It's so powerful, it has to be right."

But, he said, there's more. "You've heard about Per Bak and self-organized criticality?" I hadn't, but Stu was already telling me about it before I could answer. "It's another strand in this story. I have a feeling that all this shit links together in some wonderful way."

Per, a physicist at Brookhaven National Laboratory, New York, is at once an imposing and a jovial figure. Majestically tall, he has a round face, round spectacles—and, though apt to be absent-minded, he possesses the sharpest of intellects. Recently he developed the hypothesis that large, interactive systems—dynamical systems—naturally evolve toward a critical state. The system may be biological, like a coevolving ecosystem, or physical, as in the interaction of tectonic plates and their role in earthquakes. All this sounds a bit like the edge of chaos, I ventured. Is it? "*I* think so," he replied. "We're talking about the same kind of phenomenon."

Systems that have reached the critical state display one very characteristic property, Per explained. Perturb such a system, and you might get some small response. Perturb it again, with the same degree of disturbance, and the thing might collapse completely. Perturb it many times while poised at the critical state, and you'll get a range of responses, which can be described by a power law; that is, big responses are rare, small responses are common, and intermediate responses fall in between. "You see this with earthquakes, forest fires, Conway's Game of Life," Per explained. Would you expect to see it with extinction events, I asked, the sort of thing you see in the fossil record? "You would." And speciation events, if the environment was altered, promoting the origin of new species? "I'd expect that, too."

Per has an appealing visual analogy for a system at the critical state: a sand pile. Run a thin stream of sand onto a round plate. A pile steadily builds, soon reaching the edge. The initially low pile now gets higher and higher, until suddenly more sand may trigger a small avalanche, and then a big one, avalanches of all sizes. The sand pile, when it can take no additional sand, represents the system poised at the critical state. And the avalanches of all size ranges, provoked by disturbances of the same magnitude of disturbance (another grain of sand), represent the power law distribution of response: the signature of a system that has got itself to the critical state. Got itself, perhaps, to the edge of chaos.

If the critical state and the edge of chaos were equivalent phenomena, then an obvious question offered itself. Can you test your model ecosystems to see what happens when you perturb them? I asked Stu. "It was easy to do," he said. "We just made the external world—the abiotic world—another random connection." If the fitness landscape of one species is deformed by such external perturbation, the species is likely to become less fit. Through mutation and selection it will then reclimb the peak, or a new peak, a change that in all probability will deform the fitness landscape of one or more species with which it interacts. If connectedness among species within the system is low, then the effects of the initial perturbation will soon peter out. This is when the system is near the frozen state. With high connectedness, any single change is likely to propagate hectically throughout the system, with many large avalanches. This is the chaotic state. At the intermediate state, the edge of chaos—with internal and between-species interactions carefully tuned—some perturbations provoke small cascades of change, others trigger complete avalanches, equivalent to mass extinctions. "With our system at the edge of chaos, we saw a power law distribution of change," Stu said of the experimental test he'd run on the computer. "I don't think that's trivial. I think it's telling us something deep about the world out there."

The "something deep" is this: coevolving systems, working as complex adaptive systems, tune themselves to the point of maximum computational ability, maximum fitness, maximum evolvability. I couldn't help thinking again of "the ghost in the machine," the now-discredited phrase once used to describe an autonomous "mind" inside the brain. A ghost of sorts lives in complex adaptive systems, it seemed to me. At least, you couldn't describe evolutionary history as "one damn thing after another." If the edge of chaos is more than the seductive product of complex computer models, then the world "out there" has a thread of tantalizing inevitability to it. Perhaps more than a thread. But is it true?

I would have to try to find out.

CHAPTER FOUR

Explosions and Extinctions

Life on Earth is more than 3.8 billion years old, but only organisms built from many different kinds of cells hold a fascination for those with a passion for patterns. After such organisms evolved, some 600 million years ago, all hell broke loose, and the history of life on Earth has been one complex pattern ever since.

Within a few million years of this major turning point in Earth history, the seas were swarming with myriad forms of life, swimmers, prowlers, sedentary beasts, and burrowers. So dramatic was the event that the colloquial term Cambrian explosion is no exaggeration. Three billion years of mind-numbing biological simplicity was replaced overnight—in geological perspective—by burgeoning complexity.

Once established, multicellular life continued an upward increase in diversity, so that modern seas contain twice as many species as in the Cambrian world. That increase was no steady trend, however, with each era routinely and predictably notching up new gains in diversity. The passage of time was marked by a continuous turnover, with new species replacing existing ones. And, most dramatic of all, any steady march there might have been from ancient to modern times was interrupted by occasional catastrophic collapses in diversity, mass extinctions that in one case felled as much as 96 percent of existing species within a geological instant.

Five such events punctuate the history of life. Many lesser collapses, not big enough to deserve the appellation "mass extinction" but nevertheless devastating on a continentwide scale, also took their toll. As a result, 99.9 percent of all species that have ever lived are now extinct. As one statistical wag put it, "To a first approximation all species are extinct." Notwithstanding statistical insignificance, we and the other extant species—which number between 10 and 30 million—are the latest expression of a 600-million-year-old process of origination and extinction.

Two major patterns dominate that history. The first is its beginning, the Cambrian explosion, which is unique in several important respects. The second is the repeated collapse of biodiversity, the mass extinctions and their lesser cousins. If the new science of Complexity—with its dual notions of self-organization and the edge of chaos—is to be of any interest to biologists, it must be able to illuminate in some direct or even indirect way these two major patterns. Within the broad sweep of Earth history we are seeking the footprints of complexity, however faint.

"Hello, Roger, I think I've got the answer," enthused the voice on the other end of the telephone. "I think I can explain the pattern. It's all to do with my rugged landscapes." This was the end of July 1988, and the caller was Stu Kauffman. Earlier in the month I had written an article in *Science*, called "A Lopsided Look at Evolution," in which I described some questions posed by the Cambrian explosion, and recent ideas about their solution. "Rugged landscapes explain why you get high-level innovation in the Cambrian, but not later," continued Stu.

The key question about the nature of the Cambrian explosion relates to evolutionary innovation, not so much its quantity, which was great, but its quality, which was extraordinary. "Unprecedented and unsurpassed," was how James Valentine described it to me. Valentine, who at the time of the *Science* article was at the University of California, Santa Barbara, but is now at Berkeley, has made a long study of the Cambrian explosion and its consequences. "It's the single most spectacular phenomenon in the fossil record," he said.

True, there have been tremendous bursts of innovation later in the history of life, mostly in the wake of mass extinctions. For instance, following the Permian extinction some 250 million years ago, in which an estimated 96 percent of existing species perished, the rate of innovation almost matched that of the Cambrian. But the innovation was principally variations upon existing themes; no major new themes were added. In the Cambrian, by contrast, innovation was largely at the level of producing new themes, with variations upon them being relatively minor. "That's the challenge," said Valentine. "You have to explain the shift from few species in many groups in the Cambrian, to many species in fewer groups later."

The groups that Valentine was talking about are among the highest levels in the hierarchical structure of biological classification: classes, and, in particular, phyla. Phyla, which come just below the level of kingdom (animals and plants, to use the traditional, "commonsense" classification) and just above classes (mammals, reptiles, and so on), represent major body plans in the diversity of life. For instance, the *Arthropoda,* the most populous of all phyla, have jointed appendages, and include such creatures as insects, centipedes, millipedes, spiders, and crabs. Humans, and all the other vertebrates that so dominate our view of the world, are part of the phylum *Chordata.* Phyla are discrete body plans, upon which many variations may be created.

There are thirty major phyla in today's world, just as there have been for much the past 500 million years, a striking continuity of anatomical designs, upon which as many as 50 billion variants have come and gone. In the aftermath of the Cambrian explosion there may have been as many as a hundred phyla, the majority of which became extinct in short order, leaving the modern level of diversity. "Innovation as such wasn't unique to the Cambrian, but high-level innovation was," said Valentine, stating the problem succinctly. Tremendous evolutionary experimentation, followed by a severe sorting process, that's what the Cambrian and immediate post-Cambrian world experienced.

The discovery of this pattern makes one of the great stories in the history of paleontology. It begins essentially with Charles Darwin, ends with Harry Whittington, a professor of geology at Cambridge University, England, and has been chronicled by Stephen Jay Gould in his book *Wonderful Life*. A key player in the story was Charles Walcott, one-time secretary of the Smithsonian Institution, and discoverer in 1909 of the most spectacular window onto the Cambrian world, the Burgess Shale. The Burgess Shale deposits, which entombed an astonishing encapsulation of near-shore life in a brief moment of Cambrian history, are located high in the Rocky Mountains of British Columbia, Canada. To Walcott, the view through that window was of a world patterned just like today's, more primitive, certainly, but of equal diversity. He was being a good Darwinian.

Because Darwin viewed natural selection as an essentially gradual process, capable of producing great innovation but only incrementally over very long periods of time, he balked at the idea of rapid change. Darwin considered rapid change, whether in the appearance or disappearance of species, to be a challenge to his theory. He explained the apparently abrupt appearance of multicellular life in the Cambrian by saying that earlier forms must have existed, but are hidden in the incompleteness of the fossil record. "[D]uring these vast, yet unknown, periods of time, the world swarmed with living creatures," he wrote of the pre-Cambrian period, in *Origin of Species*.

When Walcott discovered the profusion of life forms in the Burgess Shale, which doubled what previously had been known from the Cambrian, the challenge of rapid appearance was greatly exacerbated. Precursors to Cambrian animals, which Darwin said must have existed, had not been discovered in the quantities he predicted. And the wealth of Cambrian life so apparent in the Burgess Shale made the instant of rapid appearance yet more explosive. Walcott, a devoted Darwinian, had two choices.

He could either say that the Burgess Shale discovery confirmed an explosive appearance of life in the Cambrian, and thus prove Darwin wrong. Or he could view the Burgess creatures in the context of gradualism and continuity—denying sudden appearance of great diversity—and thus maintain the Darwinian view. He chose the latter. To Walcott, and to two generations that followed him, the Cambrian not only was the product of a long history of evolution but it also presaged perfectly the present world, with its thirty or so phyla. By assigning to modern phyla virtually all the creatures he saw in the deposits, no matter how bizarre, Walcott effectively turned a blind eye to the real diversity that exploded into life in the Cambrian.

It wasn't until Harry Whittington and his students came along in the 1960s and '70s that the truly unusual nature of much of the Burgess fauna was recognized. Whittington, an expert on trilobites, discovered that Walcott had squeezed many Cambrian creatures into groups where they did not belong. Ultimately, as many as a hundred phyla were proposed, not the thirty implied by Walcott and believed by all. The Cambrian, without doubt, had now to be seen as an event of explosive evolutionary innovation. Equally, it had been followed by massive extinction, which shaped the world we know.

"There are two competing hypotheses to explain the pattern of innovation in the Cambrian," Valentine told me. "The first is the ecological hypothesis, which I tend to favor, and the second is the genomic theory. Maybe they're not strictly competing," he added. "Maybe there's something of each." The ecological theory is appealing in its simplicity. The first Cambrian organisms entered a world devoid of competitors, a world that may have swarmed with bacteria and single-celled algae, which represented more a potential source of food than a competitive challenge. With a panoply of ecological niches open to them, all kinds of evolutionary variants were viable: evolution therefore proceeded by long jumps rather than incremental creeps. "After the Permian extinction, even though countless species were lost, the full range of ecological niches would still have been occupied," suggested Valentine. "The opportunities simply weren't there, as they had been in the Cambrian." Innovation in the post-Permian was therefore much more constrained, or so the ecological hypothesis suggests.

According to the genomic hypothesis, evolution in the Cambrian produced so many experimental forms because the species' genetic packages—genomes—lacked a degree of coherence and tight control that developed later. More drastic mutation was therefore possible and viable, thus generating themes rather than variations upon themes.

The ecological hypothesis is driven essentially by external factors, namely ecological opportunity; the genomic hypothesis by internal factors, the feasibility of substantial mutation. "Neither is correct," claimed Stu Kauffman, that July day back in 1988. "Though it has more to do with genes than ecological opportunity."

Stu had been working on his notion of rugged fitness landscapes, but had not reached the ideas about coevolving landscapes and the edge of chaos. That came later. Mutations that affect embryological development early in the process can produce dramatic alteration in the adult form, because the initial small change becomes magnified as development unfolds, Stu explained. "Embryological development represents a rugged fitness landscape, on which you very rapidly reach local optima." Which means, I ventured, that reaching fitter variants early in development becomes more and more difficult? "That's right. Early ontogeny gets frozen in, and new variants increasingly have to come from late in development, and these produce less dramatic evolutionary shifts. The Cambrian creatures could exploit new fitness variants early in ontogeny, and get big evolutionary leaps, but by the time the Permian came along, that game was over, and only small changes were possible." Stu wrote up his ideas, published them in the journal *Evolutionary Ecology*, and sent me a copy. I thought nothing more about it. Until Complexity came along.

I asked Stu recently whether he thought that the rugged landscape explanation was still valid. "Yes, as far as it went," he replied. "But I think it should be looked at in a coevolutionary context, which I didn't do before." In that case, I said, how would you apply your coevolution model of the edge of chaos here? There was a pause, uncharacteristic and long enough for me to realize that we were entering uncharted territory. I had thought about the Cambrian explosion in qualitative terms, at its simplest a shift from single-celled to multicellular organisms. I'd thought about the kind of ecological interactions that must have developed among populations of organisms in the pre-cambrian world, mixed populations of bacteria and algae in stromatolites, for instance. And in an abstract sense, these complex mini-ecologies became interactions within a single organism in the post-cambrian world: a hierarchical shift had occurred. But what about the edge of chaos?

Is it reasonable, I asked Stu, to think about the Cambrian animals as part of a system in the chaotic regime, a system that was moving toward, but not yet reached, the edge of chaos? "That would make sense," he said. "In my model they would have to be interacting with each other quite closely. If they had recently evolved from a few common ancestors, that'd be reasonable." If the Cambrian explosion truly represented a chaotic regime, perturbations would cause big avalanches of change, according to Stu's model. And this would include a greater propensity to evolutionary innovation, perhaps producing unusually innovative novelties. As the system coevolved to a balanced state (the edge of chaos, in Stu's coevolution model), responses to perturbation would diminish, with innovation becoming less adventurous, until a steady turnover state was reached. "All you would need to explain the difference of innovation after the Permian extinction is that the system is pushed again into the chaotic regime, but not as far," Stu suggested. "Innovation would occur in the postextinction rebound, but would be less exaggerated."

Plausible, but complete speculation. How could we know? I asked. What could we look for as a clue? This time there was no hesitation. "Look at the extinctions," Stu instantly responded. "If there's any merit in this crazy idea, then extinction rates in the Cambrian would have been higher, too, and that would mean species' lifetimes would have been shorter." And in other periods after mass extinctions, I added. "Yes, that's right. What's the evidence?" We felt we might be close to something new and interesting. I don't know, I replied. I'll find out. The result? I asked paleontologists who would know, and repeatedly found the same answer: no reliable data.

I retreated from my foray into the process of innovation in the Cambrian, and addressed two other questions. The first concerns its overall pattern of innovation. Is the pattern unique, or is there something more general here, something that may illuminate in a fundamental way the process of innovation in complex systems? The second question asks whether the products of that process of innovation are unique. If we could turn the clock back and run the Cambrian explosion again, would the world look as it does today? Would humans be here to observe it and think about it?

When George Gumerman read a review of the Burgess Shale fauna and the Cambrian explosion in the 20 October 1989 issue of *Science*, he found enough similarities in the dynamics of the system to make him look again at his own area of interest: the evolution of Southwest prehistoric societies. As one of the co-organizers of the Santa Fe Institute's conference on social complexity in the Southwest, George had thought the pattern of innovation pertinent. He pointed out that great diversification in social conventions occurred in the Southwest in the century and a half from A.D. 1000 to 1150; and this was followed by an equally dramatic reduction in diversity, with the Chaco Canyon culture dominant. Experimentation followed by specialization; it's the pattern of life in the Cambrian, and of life for the Anasazi in the American Southwest. Are there other examples?

"You see it with technological innovation in industrial societies, too," Stu told my subgroup at the Southwest conference. "Think of the first bicycles or the first cars," he continued. "Lots of experimentation to begin with, different forms of bicycle, different forms of propulsion and design for cars, all viable. As time goes on and the world gets full of cycles or cars, or whatever it is you're thinking of, the extremes get weeded out, a few forms survive, and subsequent innovation focusses on improvement on the remaining themes. You go from generation of many themes to variations upon a few, just like the Cambrian."

With innovation in the human realm, however, there is always the possibility of reviving a vanished design. We could, for motives of fashion or economic necessity, resurrect the penny-farthing bicycle or the steam-driven automobile. For evolutionary history, however, lost forms are just that—history. Extinction is forever, and the reevolution of extinct major forms requires the concatenation of too many improbable events for it to occur. The only trilobites and dinosaurs we will ever encounter are those in books, museum collections, or in geological exposures. Historical contingency therefore influenced the shape of the Cambrian explosion and its aftermath. Phyla that were lost, whether through the exigencies of competition or stochastic elimination (bad luck, in other words), were never reinvented. The shape of today's world was, it seems, influenced to a large extent by which phyla survived 500 million years ago.

What of social and cultural traditions that go extinct through periods of experimentation, such as occurred among the Anasazi almost a millennium ago? Is their fate likely to follow that of steam-driven automobiles or trilobites? "You can extend the pattern of Cambrian history to human cultural history," George told me. "We've seen how the pattern of innovation is similar, with a burst of novelties and then subsequent loss. And just as you don't get extinct forms of animals reappearing, you don't see the exact reappearance of cultures once they've changed." There is so much historical content to cultural traditions and mythologies that, once lost, they are unlikely to be exactly reformulated, George explained. In other words, cultures are more like trilobites than automobiles.

So, how consistent a pattern is there in innovation in complex adaptive systems such as these? Each is affected by historical contingency, but to different degrees. It is surely significant that, with all these differences of detail—in the biological, cultural, and technological realms—the overall pattern is remarkably similar. It encourages the belief that consistency of pattern is more than mere coincidence or mere analogy. Fundamental dynamics may be at work, making the pattern of innovation in complex adaptive systems predictable to a degree.

In his *Science* article on the Cambrian explosion, Simon Conway Morris concluded with a small thought experiment. "What if the Cambrian explosion was to be rerun?" he asked. The same explosive innovation would occur, he ventured, almost certainly with the same trimming of diversity in the aftermath of massive innovation. Subsequent fauna would occupy niches instantly recognizable to our eyes: herbivores, carnivores, insectivores, and so on—creatures large and small, making livings such as those in the past and those today. But, he suggests, because of historical contingency, the creatures themselves would look like nothing we've experienced, and would be "worthy of the finest science fiction."

The fact of historical contingency, which Stephen Jay Gould has championed ever more strongly in recent years, means that the world we inhabit is simply one of a virtual infinity of worlds. Run the tape again, he says, and even the most modest turn on the long road of history translates into dramatic effect a hundred million years or so later. Multiply such small excursions of fate a millionfold, and the end result is a world unrecognizable to our eyes. Or is it? Is there a virtual infinity of possible worlds, of which our experience is just one? Not according to Brian Goodwin.

For Brian, the mechanics of embryological development are tightly constrained, and this greatly limits the kinds of structures and the kinds of species that can arise. In the language of complex dynamical systems, the space of morphological possibilities is thinly populated by attractors, those states to which dynamical systems eventually settle, ghost species that might be brought to life under the correct circumstances. This image is very different from the standard outlook of Darwinian evolution, in which the processes of natural selection and adaptation can explore virtually any and every corner of that space. I asked Brian if he would argue that in his world only a limited range of species was possible, whereas in the adaptationist's world there was an infinity of possible species. "That overstates the case a little," Brian responded. "Even the most ardent adaptationist allows for some constraints on morphological possibilities, such as basic biomechanics. Few people would argue that terrestrial organisms might evolve wheels, for instance. But overall, your comparison is correct."

To most biologists, Brian's dynamical systems approach is barely comprehensible at best, completely crazy at worst. "A plausible conjecture," Brian reminded me again. OK, but let's push it a little further, I responded. You're saying that the world out there is populated by a range of ghost species, dynamical attractors, only some of which may be occupied at any time? "Yes, that's a fair statement. You shouldn't see the attractors as static, however. They will change, dynamical possibilities will change, as the environment changes. I've been guilty in the past of ignoring the effect of the environment, but it is important. It may diminish the stability of some attractors, and improve the stability of others."

An obvious question about historical contingency thrust itself forward. If it's true that only a limited number of attractors populate potential morphological space, I ventured, does that mean that if you reran the Cambrian explosion, the new world wouldn't look so very different from the one we know? That Steve Gould might not be correct when he says: "The divine tape player holds a million scenarios, each perfectly sensible."?

"Let me answer the question this way," Brian began thoughtfully. "You are aware of the phenomenon of convergence in biology, when you see strikingly similar morphology in widely divergent species?" Like the Tasmanian wolf and the true wolf, I offered. "Yes, one's a marsupial, the other a true mammal, separated by, I don't know, 50 million years. And yet anatomically they are virtually the same." Conventional evolutionary theory explains the phenomenon partly as historical contingency: both are derived from the same mammalian ancestor. But the nub of the argument is that similar adaptations shape similar anatomies and behaviors. "That, I believe, is stretching a weak argument to breaking point," said Brian. "No two environments are so similar as to produce such parallel anatomy."

So you would say that the Tasmanian wolf and the true wolf are attractors in the space of morphogenetic rules? "Yes." And you would extend the argument to a rerun of the Cambrian explosion? "I'd extend it part of the way," said Brain. "Let me give you an analogy. "Suppose you reran the Big Bang. What are the chances of getting the same periodic table of natural elements, the same ninety-two combinations of protons, neutrons, and electrons? Pretty good, or so I'm led to believe. I think of a rerun of the Cambrian explosion in the same way, not to the same extent perhaps, but as an image. If there are dynamical attractors in the space of morphological possibilities, as I believe, then a rerun of the Cambrian explosion would produce a world much more like the one we know than Steve Gould says. It wouldn't be identical to the one we know, but there may be a lot of similarities, ghosts we'd instantly recognize."

In other words, evolutionary history may not be one damn thing after another, but would to an interesting extent be inevitable. By now this is becoming something of a refrain for complex adaptive systems.

"The study of extinctions has never been fashionable," lamented David Raup. "There are many reasons, I'm sure. But you can probably trace a lot of it back to Darwin." Dave is a geologist at the University of Chicago, where the Department of Geophysical Sciences sits on South Ellis amid the faux Gothic of much of the university's architecture. Geophysical Sciences, however, is a modern building, all red brick and glass, windows at inventive angles, that kind of thing. Dave's office is large, sparse, illuminated by two of those windows, and, for the domain of a leading scholar in his discipline, surprisingly free of rocks and fossils. Despite his rugged outdoors appearance, Dave's research milieu is more the computer and high-powered statistical analysis rather than hacking rocks out of ancient strata. "Being wedded to gradualism, Darwin tried to deny the existence of mass extinctions," explained Dave.

Just as he tried to deny the abrupt appearance of multicellular animals in the Cambrian, I suggested. "That kind of thing, yes. Darwin said that extinction was 'the most gratuitous mystery,' something like that. But he also said that the origin of species is gradual, and so is their extinction." A sentence from *Origin of Species* captures this: "Species and groups of species gradually disappear, one after the other, first from one spot, then from another, and finally from the world."

If species disappear as part of the gradualistic dynamics of natural selection, then there is nothing more to explain. There is nothing more to be said about the process, except the necessity to label as failures those species that become extinct. Darwin put this explicitly: "The inhabitants of each successive period in the world's history have beaten their predecessors in the race for life, and are, insofar, higher in the scale of nature." (Everybody knows that the dinosaurs were failures, don't they?) Darwin's reluctance to contemplate the abrupt disappearance of species rested not only on the gradualistic mode of natural selection but also on the notion of gradualism in geology as a whole, which had been promulgated by his friend Charles Lyell.

At the beginning of the nineteenth century, the great French geologist and naturalist Baron Georges Cuvier proposed what came to be known as the Catastrophe theory, or Catastrophism. According to the theory, the abrupt faunal changes geologists saw in rock strata were the result of periodic devastations that wiped out all or most extant species, each successive period being repopulated with new kinds of animals and plants, by God's hand. Lyell rejected so nonscientific a hypothesis (as did James Hutton before him), and replaced it with the notion that geological processes proceeded gradually—*all* geological processes. In his major work, *Principles of Geology*, Lyell said that abrupt transitions in the geological record would one day be shown to be erroneous, when transitional strata were discovered. Gradualism superseded Catastrophism. Darwin's and Lyell's worldviews were therefore perfectly complementary.

"Eventually, geologists came to see the apparent abrupt changes in the geological record as real, as mass extinctions, but Darwinian explanations—competition, predation, and so on—lingered and dominated what little debate there was," said Dave. Intellectual excitement over extinction mechanisms remained conspicuously absent in the biology community. Then came Luis Alvarez. "Outrageous in his behavior, outrageous in his suggestions, he really stirred things up," Dave said mischievously. Alvarez, a physicist at the University of California, Berkeley, in 1980 suggested with several colleagues that the mass extinction 65 million years ago, which saw the end of the dinosaurs, was caused by Earth's impact with a giant asteroid. Dave loved the idea, partly for its audacity but also for its plausibility. Four years later, Dave and his colleague Jack Sepkoski went a step—or several steps—further than Alvarez and suggested that major asteroid impacts occurred every 26 million years, causing periodic mass extinctions.

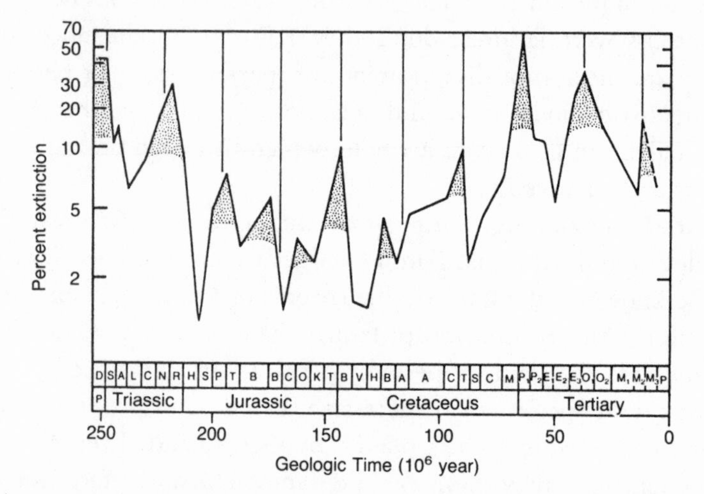

Fig. 4. The history of life is punctuated by mass extinctions. The spikes in the graph indicate periods of high extinction, with the percentage of families going extinct shown on the left-hand side. Courtesy of David Raup and John Sepkoski.

These days most people in the field accept that the end-Cretaceous extinction was caused by asteroid impact. The same explanation is considered possible for a small handful of other extinctions. But the notion of periodic impact, every 26 million years, remains distinctly controversial, not least because it smacks too much of the old idea of Catastrophism for most people's taste. Dave has persisted, however, and recently drew together evidence on the timing of asteroid impact (from the age of large craters) and the timing of mass extinctions (from the fossil record), and showed they match closely. "As much as 60 percent of all extinctions may have been the result of asteroid impact," he concluded. Sixty percent. That's enormous, I said, not hiding my incredulity. "I know. People are going to have a hard time with it, but it's a credible hypothesis."

Implicit in all proposed causes of mass extinction, including asteroid impact, is an equality between cause and effect, between the scale of environmental perturbation and the proportion of species dying off. That's how a world governed by linear equations would work: big perturbations produce big extinctions, small perturbations, small extinctions. "That's not necessarily true," said Stu Kauffman, "not if we're right about ecosystems being poised at the edge of chaos." The world Stu was describing was a nonlinear world, where complex dynamics of the sort we've already encountered produce complex patterns. "It may be that similar changes in the environment can produce extinctions of all magnitudes," Stu added.

We were talking in his office at the University of Pennsylvania, white-coated biochemists dispensing precious liquids with exquisite accuracy in the lab next door, excursions into experimental molecular evolution, while theories of global extinctions were discussed in here. Seemed bizarre. You're talking about your coevolutionary model? "I am," said Stu. "You'll remember when our model ecosystem got itself to the balanced state, the edge of chaos, we tweaked it with some kind of external change, and produced avalanches of change of all sizes?" I did. Like avalanches on Per Bak's sand pile, poised at the critical state, giving a power law distribution? "Exactly. That was part of our reasoning that our model ecosystem had come to the edge of chaos."

And what you want to know is whether real ecosystems—out there—are also poised at the edge of chaos? "Maybe nature has done the experiment to give us the answer," responded Stu. "Maybe the answer's in the mass extinction data." Stu and his colleague Sonke Johnsen had gotten the data on mass extinctions (from a paper of Dave Raup's), and plotted the magnitude of extinctions against their frequency, and tested for a power law. "You get something that's very close to a power law," said Stu. "It's not a straight line, just slightly convex, but that's what you see with extinctions in our model ecosystem." If the data from your coevolving ecosystem is the same as you get with Dave's data from real extinctions, that means that the world out there is poised at the edge of chaos? "That's how I'd interpret it," said Stu. "Just slightly on the frozen side of the edge of chaos." The ecosystems, model and real, are just on the frozen side of chaos, Stu suggested, as a result of continuous perturbation.

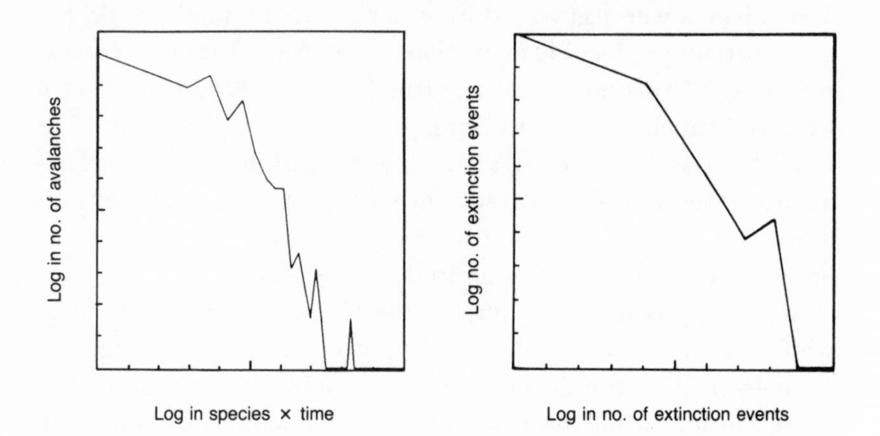

Fig. 5. Extinctions in model and real ecosystems come close to a power law distribution. The graph at left shows a plot of the log of the number of avalanches (equivalent to extinctions) on the vertical axis against the log of the size of the avalanches in Kauffman's model ecosystem, with 100 species (both axes plotted as natural logarithms). A perfect power law distribution would be a straight line, downward slope from left to right. The plot on the right shows the same thing for extinctions in the fossil record. The slopes are similar, and both are slightly on the frozen side of the edge of chaos. Courtesy of S. A. Kauffman and S. Johnsen.

"If we're right, we not only know something about the dynamics of ecosystems in the real world in the context of complex adaptive systems," continued Stu, "but we are also faced with the counterintuitive notion that a mass extinction like the end-Permian could have been caused by the same kind of perturbation that produced a small blip on the extinction map." If you're right. "Yes."

Dave shook his head slowly when I asked him if he thought the mass extinction data showed a power law distribution. He's more thinker than talker. At length he said, "I don't think so," the "I" stretched out as if over half a dozen syllables, doubt dripping from each one. "First of all," he continued, "the data are lousy." But they're your data, I said, surprised. "I know that, and I don't like going around saying my data are lousy. They're the best we have, but you'd like something a whole lot better if you're going to draw conclusions of this magnitude." The data in question record measures of extinction levels at seventy-nine points during post-Cambrian life. Dave and his colleague Jack Sepkoski have spent years compiling these data, sometimes together, sometimes independently. It's a time-consuming, very difficult business, and the quality of the end product reflects the problems in compiling them, not the scholarship involved. "Someday we'll have good enough data to do this sort of thing."

Just suppose, I said, the data were good now, and you saw a curve like the one Stu got, something close to a power law. Would his interpretation be valid? "I simulate for a living," he began by way of reply, "and I know how delicate a process it is, how fraught with traps. You find something that reflects the real world in some way and you think, 'Hey, I'm on to something here.' It's a gestalt thing, and it's seductive." He told me about a snowy winter weekend that he spent at the Woods Hole Marine Biological Laboratory in Massachusetts, many years ago, with Steve Gould, Dan Simberloff (an ecologist), Jack Sepkoski, and a few others, brainstorming with the hope of coming up with something insightful into the problems of evolutionary biology. Nothing was coming of it. Then on Sunday afternoon Dave suggested they look at some of these processes as if they were random. The suggestion spawned about a dozen research papers in the years to follow. Some good, one really bad.

"Steve, Dan, and I did a simulation in which we assigned equal probabilities of extinction to a model biota, then looked at extinction patterns," Dave told me. "Lo and behold, we got patterns just like you see in the fossil record: groups waxing and waning, some going extinct finally, just like the real world." This was at a time when a number of researchers were trying to break away from the "everything is determined by the inexorable dynamics of natural selection" theme of evolutionary biology. It was a big surprise that major extinction patterns could be produced using a purely random approach. It was also wrong. "We blew it on the scaling," Dave explained. "Our groups were much too small, as Steve Stanley rightly pointed out. If you use much larger groups, you don't get that pattern at all. So, you see, I'm familiar with how easy it is to be led astray, especially by something that looks so compelling."

Is Stu being led astray? I asked. "There's something called the broken stick model in statistics," said Dave. In this trick, a random number generator "breaks" a stick a hundred inches long at twenty-five points, and produces twenty-six short sticks. Measure them, count the number that are one inch long, the number two inches long, and so on, and draw a histogram. You get a skewed distribution, toward the short end, just like many natural phenomena, including the distribution of sizes of U.S. cities, for instance. "One thing you have to remember about extinctions is that some species are more likely than others to die out, just because they exist as small, isolated populations," explained Dave. "This sort of statistical quirk can skew your results, easily." So, you would be suspicious of anything that looks like a power law? "I would, because it's so common, just in the nature of statistics. It may tell you that a system is poised at a critical point, whatever that means, but it may not. In any case, when Stu says that the curve he gets from my data is close to a power law, he knows that there are many other mathematical models that could fit equally well."

Clearly, there were many reasons to be cautious about drawing the conclusion that global ecosystems are poised at the edge of chaos, using just the extinction data. "You asked me how I would interpret a valid power law distribution for extinction sizes," said Dave, returning to the original question. "Well, you know I argue that a very large fraction of extinctions may be caused by asteroid impact. And you know that the size of asteroids can be described by a power law: big ones are rare, small ones are common; you can see that from the size distribution of craters on the moon. So it could be that a power law distribution of the size and frequency of extinction is a reflection of the power law distribution of the size and frequency of asteroid impact. Couldn't it?"

"Yes it could," conceded Stu when I put Dave's question to him. So how would you know whether the power law distribution of extinctions is caused by the size distribution of asteroids or the fact that global ecosystems are poised near the edge of chaos? "I asked Per [Bak] what happens when you get one power law imposed on another, what would it look like?" said Stu. "He said, 'You'd still see a power law.' I kept asking him how, but didn't get anywhere. It'd be pretty messy, I think." Stu also conceded that the curve from the extinction data could be described by other mathematical models, not just the power law. "But at least the curve is consistent with the global ecosystems being near the edge of chaos," he said. "Yes, I know that's pretty weak. But, look, if the curve looked nothing like a power law we wouldn't have anything to talk about. The edge of chaos wouldn't be in it. As things are, it remains possible that global ecosystems bring themselves to the edge of chaos, as predicted by our models."

Finally, I asked Stu about connectedness. The coevolutionary model builds in connections between species in the ecosystem. It's part of the system, the part that gets tuned as the system moves itself toward the edge of chaos. Connectedness is required if the ecosystem is to work as a whole, not just as independent entities. And connectedness is required if perturbations are to cascade through the system, producing avalanches of speciation and extinction. In a brief excursion into the biology of the problem, Dave had expressed some doubt that connectedness in the real world would be sufficient to propagate the consequences of perturbations world-wide, and thus cause mass extinctions on a global scale. Is Dave correct in being skeptical? I asked. "It gives one pause," replied Stu. "I don't have problems thinking about connectedness on a continentwide scale, but globally, that's a challenge. You'd need Gaia-like connections."

As Stu was pondering this problem, he said, "Yes, getting a 96 percent hit, like the end-Permian extinction, that would need *a lot* of connectedness." Wait a minute, I said, you know that the Permian extinction coincided with the coalescence of the continents, to form Pangea? "Did it? Hey, that's great."

The world's continents are in constant, barely perceptible motion, passengers on a thin crust that's divided into many so-called plates. One of the great discoveries of twentieth-century science, the fact of continental movement as a result of plate tectonics, puts history in a new light, one that is very difficult for human minds, so much in the thrall of the present, to understand. Engaged in a global shuffle, the continents occasionally jostle each other, and occasionally coalesce as a single supercontinent, Pangea. The last time this happened was at the end of the Permian, 250 million years ago. "That would mean that all biotas would be in potential contact with each other, all of them," said Stu.

The collisions and coalescence of all the continents has indeed been invoked as contributing to, if not causing, the end-Permian extinction. The reasoning is that as you bring land masses together to form one giant continent, about half the coastline is lost. (Make four one-inch-square pieces of card, measure their total edge length; now bring them together as a large, single square, and measure them again; then you'll see.) Extensive extinction in the marine realm is likely from this fact alone. "That's true, of course," said Stu. "But it also gives you the potential connectedness for coevolutionary avalanches across the entire land mass, to contribute to the biggest extinction in Earth history, doesn't it?" We were deep in speculative territory here, and any footprints of complexity we might spot would have to be viewed with a high order of skepticism.

Footprints I had seen elsewhere, however. Indistinct and questionable, it's true. Nothing to justify the declaration that, in the Cambrian explosion and mass extinctions, complexity triumphs as a dominant force. But sufficient to encourage further exploration of patterns in biology. I knew I needed to come down from the heights of the larger patterns and look more closely at ecosystems themselves, both real and those that live only in computers.

CHAPTER FIVE

Life in a Computer

"You can't spend any time in a rain forest and not be enthralled by it," said Tom Ray. "I was overwhelmed by the experience when I first came here, eighteen years ago. I still am." We were dressed for the heat and the wetness: thin cotton pants and shirts, bush hats, rubber boots, and, wonderfully incongruous it seemed to me, umbrellas. Incongruous or not, they proved their worth. "This *is* a *rain* forest," said Tom, greatly amused at my concern for sartorial correctness over comfort. We were deep in La Selva Biological Reserve in north-central Costa Rica, part of this small country's extensive protected primary forest system. It was January, which passes for the dry season in these parts. "You should be here when it *really* rains."

I consider myself well travelled, privileged to have visited some of the more ecologically exotic parts of the globe, including the East African savannah, the high Andes, and the Galápagos Islands. This was my first rain forest. Unprepared, that's the best description of my state. Unprepared for how very open it all was, as you walk among the exquisitely buttressed trunks of giant trees beneath the high canopy, a modest tangle of vegetation covers the ground. (Dense thickets at ground level occur only in regenerating, secondary forest, Tom explained.) Unprepared for how quiet it all was. (It was midmorning, and countless birds, their dawn chorus over, would be silent until twilight; likewise the howler monkeys.) And unprepared for the diversity of life. "Every niche teeming with life," runs the cliché. And it's true. "More species per acre than anywhere on Earth," said Tom. "Look around; you'll see more species of tree in this small space than in an entire temperate forest."

The number of tree species was just the beginning. Each tree was host to another level of diversity, festooned as they were with epiphytes in every crevice in the trunk and perched on every secure surface along branches: night-blooming cactus, orchids, ferns, bromeliads, aroids, as well as lichens, mosses, and liverworts. Vines hung everywhere. Pictures cannot prepare one for this reality. Bewildering novelty to me, all this was familiar territory to Tom, who was constantly looking *for* things as I was simply looking *at* them. The utility of an umbrella as a probe as well as essential protection became apparent.

Tom was station manager of the reserve for a year in the late seventies and has visited the region every year since then. He has a home nearby, hidden away on forty acres of primary forest that he bought in 1982 to prevent it being turned into cattle pasture. He fought political battles, sometimes in real danger to his life, for the preservation of other tracts of rain forest. Once, Murray Gell-Mann, an ardent ornithologist, aided one of Tom's preservation efforts by urging the MacArthur Foundation to put up a million dollars to buy land. The two men did not meet at that time.

Every December Tom leaves the University of Delaware, where he is on the faculty as an ecologist, and comes to the rain forest, where he stays for a month. He is more at ease here than in the city. His warnings to me about bullet ants and the fangs of the fer-de-lance at the outset of our foray into the forest were a reminder that each environment has its hazards.

"You don't have to be an ecologist to get a sense of the complexity here," said Tom. "It's more than a richness of species, more than lots of different kinds of organisms coexisting in creative profusion. You get a sense of how the forest works as a whole." We were picking our way along a narrow track, mud sucking at our boots at each step, the vegetation illuminated by a diffuse, filtered light. I was asking about biological complexity, about patterns in ecosystems. Tom warned against the fuzzy notion of "the wonderful balance of nature" that once was so pervasive, in which everything works for the good of the community, everything structured as it should be. You still hear that kind of sentiment on some natural history shows on television, I said. "Yes, it's unfortunate," Tom replied. "Nevertheless, there *is* pattern here, on all kinds of scales, both in time and space. And pattern is what biologists should be interested in."

Suddenly he stopped. "Look." He pointed ahead. I couldn't see what he'd indicated. "Near the rubber tree." Winding across the path, drier here, was the tail-end of a column of army ants, inexorable in its progress. "No, here," said Tom, directing my attention from the ants to a scattering of white splotches on the ground, near to the column. "We should see the butterflies, the ant butterflies." Tom explained that not long after he first came to La Selva, in 1974, he discovered the previously unknown phenomenon of ant butterflies.

Army ants are notorious for their voracity as their columns surge unstoppably through the undergrowth. Also well known are the many species of birds that exploit the effects of the advancing column, namely swarms of insects that are flushed from the foliage. As the birds hover and swoop, feeding on the newly available food, their droppings mark the path of the column. Rich in nitrogen, the droppings provide nutrients for at least three species of butterfly, particularly the females, who must exploit the resource while it is still moist. Ants, followed by birds, followed by butterflies. "Look, there they are," said Tom, indicating a cluster of small yellow-orange-and-black tiger-striped butterflies, which swooped down to the droppings and equally quickly flew away, avoiding the present danger of the ants. "It's a nice example of connectedness, isn't it?" said Tom, still pleased with his discovery a decade later, still taking pleasure in the biological complexity.

Tom is a naturalist in the Darwinian tradition, a close observer of nature. He is passionate about evolution as the underlying unity of it all. "But, you know, a few years ago I was beginning to be dissatisfied, intellectually restless," he explained as we sat at the base of an ancient Gavilan tree. "I wanted to study evolution, but something was missing. All I could do was study the products of evolution—all of this," he said, with a sweep of the hand. "That's why I developed Tierra. And now I'm a naturalist in a different kind of world, an alien world—it's life in a computer."

On 3 January 1990, against all the predictions of the experts and his own expectations, Tom unleashed evolution in a computer. A simple ancestral "organism"—a small, eighty-instruction computer program—reproduced, mutated, and evolved into a diversity of descendants reminiscent of the rain-forest ecosystem that had been Tom's research milieu for so long. An E-mail message from Tom to Chris Langton at the Santa Fe Institute read: "An ecology has emerged." With that message, Tom's life changed. He still goes to the rain forest. But the evolution he studies is in his computer, a virtual world of his own creation. For the Santa Fe Institute, Tom's adventure has provided a vital bridge between abstract theory of dynamical systems and the real world of nature.

"I remember clearly when I conceived the idea of evolution in a computer," Tom told me. "It all came in a rush of ideas, complete, everything I wanted to do. But that was more than ten years ago." Some while after our visit to the rain forest I went up to Newark, Delaware, to see Tom's virtual world at first hand. His office in the department of biology was huge, with high ceilings, and lit on two sides by long windows. Two long tables occupied the middle of the room, one with three computers and a printer, the other a scatter of papers and books. The bookcase at the end of the room held dozens of computer manuals, and a copy of *Origin of Species*. On one wall hung a large poster, showing a spiral galaxy, with the word CREATION written below. On another hung a film poster, from the thirties era, of *The Jungle Princess*, starring Dorothy Lamour and Ray Milland. Different from La Selva, but with recognizable echoes. "Lots of things held me back," continued Tom, "not least of which was a naïveté about computers and programming."

Before joining the faculty at the University of Delaware, Tom had been in graduate school at Harvard, spending some of the time as Edward O. Wilson's field assistant. One evening he was visiting the Harvard Science Center, where the Cambridge Go Club met regularly. Go, an ancient Chinese game, is exceedingly complex and involves moving populations of "pebbles" around a board, the aim being to trap and destroy the opponent. Because of a certain intellectual affinity, many of the club's members were from the Artificial Intelligence lab at the Massachusetts Institute of Technology. "That evening, there was one guy playing by himself, so I sat down and he explained the game to me," said Tom.

The lone player described the game in very lifelike metaphors such as the strategy of certain groups of pebbles, the pebbles being surrounded and killed, and so on. This intrigued Tom, because it had the aura of an artificial world. Then the player casually asked a question, one that seemed to crystallize in Tom's mind a clear and powerful goal from a series of half-formed, barely conscious ideas already lingering there. "Did you know it was possible to write a computer program that can self-replicate?" the player asked. "I remember immediately the flood of ideas, all the kinds of ideas I'm pursuing now," said Tom. "I asked him how it was done and he said, 'It's trivial.' I pushed him some more, and either he didn't explain it well or I simply couldn't understand. Anyway, there I was, left with my fantasies, but no way of realizing them."

The next ten years were productive, but ultimately frustrating. Productive, because Tom's field studies of a group of vines—the *Monstera*—produced some fascinating discoveries. Not only do these plants sometimes grow toward the dark, a distinctly unplantlike behavior; but they also change their form dramatically, depending on whether they are growing on the ground, in the lower parts of a tree support, high up in the tree, or hanging down. "I was interested in morphology," Tom explained. "Morphology is the trail left by development. And ultimately, we will have to understand development if we are going to understand evolution." Important though all this was, Tom's colleagues were very traditional in their approach as ecologists, and were less than sympathetic to his work. Tom is also something of a loner, capable of intense focus on the challenge at hand, and its pursuit wherever it may take him. He didn't need, or respect, the company of his colleagues, and they knew it. "My tenure decision was coming up, and frankly the prospects weren't good," Tom recounted, still bitter. "The dean suggested I withdraw my application because, although I had the support of the rest of the faculty, the ecologists in my department were very negative."

The turning point for Tom came in 1987, when he bought his first personal computer, a modest laptop. It would open his eyes to the world of computers, and spark the notion that maybe the time had come to create evolution in a bottle, which is how he characterized it. "I had worked on mainframe computers for a long time," said Tom. "But there is a literal and figurative distance between you and the computer. You type your stuff in, get the answers back, and you don't know what's going on." The modest laptop opened a window on to what goes on in the guts of a computer. "I had been in Costa Rica for a semester, which is why I needed the laptop, and when I got back I started reading. I bought Borland's Turbo C compiler and their debugger. With the debugger I could 'see' inside the machine. I could see the memory and the central processing unit. I could see the programs in there, and how they worked on data." Why, I asked, was all that so important? "Because I had a clear sense of the computer as an environment, an environment in which my 'creatures' might evolve. It was an epiphany."

A second crack at tenure was looming, the last chance. "If you don't get tenure at Delaware, you don't get tenure anywhere," said Tom dryly. But, with all interest in "real" ecology now evaporated, and an ever-deepening obsession with creating life in a computer, making good on that last chance was going to be difficult. While going through the motions of being a loyal faculty member pursuing real ecology, Tom immersed himself further in computer books, learning to write code. He also decided he must find out what, if anything, others had already achieved with self-replicating computer programs. Computer viruses had appeared on the scene by this time, and these had at least some of the elements Tom was seeking. "I put out a message on E-mail, headed, 'Satanic viruses: blasphemy against computer cult.' The Salman Rushdie affair had just happened," said Tom. In the E-mail notice, Tom asked for information about a recent book on computer viruses, and said, "I am an evolutionary biologist and am interested in studying self-replicating code with mutation and recombination as a model for molecular evolution."

In the tradition of E-mail correspondence, Tom got back a lot of satirical messages continuing the reference to computer cult, and some serious ones. One said: "Writing a self-modifying program is still in the realm of science fiction." Not very encouraging, I ventured. "No, it wasn't," said Tom. "But I'm stubborn about these things, and I persevered." The one positive thing that came out of the foray into E-mail was a reference to a book with the title *Artificial Life*. "I knew that's what I needed," said Tom. "This was at the beginning of 1989, and the book was dated 1989. I rushed out to get it." The book was the collected papers of the First Artificial Life Conference, which Chris Langton had organized at Los Alamos in September 1987. Chris was the book's editor, and he wrote a general introduction, which explained his view of artificial life, and its prospects.

Tom opened the book, with a mixture of excitement and foreboding. "Chris spelled out beautifully the kind of research program I had in mind. Here I was, having come up with the same kind of notion independently. But then I thought, 'Oh shit, I guess it's been done.' " It's a big book, not quickly digested. Tom worked his way through it steadily, with the growing realization that, no, he hadn't been preempted. No one had done what he planned. "They seemed to have the same goal, but they had very different programs," said Tom. "Cellular automata, that's about as close as it got. The Game of Life, that kind of thing." Cellular automata are impressive in the patterns they produce and the uncanny lifelike feel they convey, but Tom was interested in programs that evolve through mutation and compete with each other, as organisms do in the real world. Nothing in *Artificial Life* came close to that.

Tom immediately sent an E-mail message to Chris, and a sometimes turbulent correspondence was established, which lasted almost a year. Tom explained his goals, and said he believed they were unmet by others in the field. Chris, excited to hear at last from a real biologist, suggested Tom might go out to Los Alamos, to visit the nonlinear dynamical systems group. So far so good. Then Tom sent Chris a copy of his essay "Artificial Life: an Ecological Approach." An aggressive but definitely well-intentioned editor, Chris effectively tore the essay to shreds, saying in summary: "Ray underestimates some of the inherent difficulty of the problems. Ignores (or is unaware of) dangers of breeding code which can survive in commonly used operating systems on the network. Overestimates differences between his approach and other work in artificial life." The "upside" comment, "On the whole a lot of good ideas, suggestions, and insights," did little to assuage Tom's hurt. "I'm sorry you were so turned off by my essay," he snapped back, the day he received Chris's review. "Whoa! I never said my reaction was negative!" replied Chris. And so it went on, the gap gradually narrowing. "I guess I was a bit defensive," Tom told me. "But I also don't think Chris fully understood what I wanted to do. To be fair, I didn't know at that point what I would do either."

Meanwhile, Tom organized a seminar for his fellow ecologists at Delaware, and gave a presentation on the concept of artificial life. "You know, they laughed. They just laughed me out of the room." Your colleagues, who were about to vote on your last crack at tenure, I said. Didn't sound good. "No it didn't. Then I got the invitation to visit Los Alamos, and it came from Doyne Farmer," recounted Tom. "Everybody had heard of Doyne Farmer. He'd been written up in Gleick's *Chaos*, and people had a lot of respect for him. Suddenly people didn't think I was so crazy." Tom, like many scientists at least part politician, knew the invitation from Farmer was imminent when he had arranged the seminar.

At the beginning of October 1989, Tom made his trip to Los Alamos, and ascended to the intellectual stratosphere. Chris Langton, Doyne Farmer, Walter Fontana, Stephanie Forrest, Steen Rassmussen—these were big names, the top people in dynamical systems, in artificial life. Their message was threefold. Tom had to attend to security, to make sure whatever he might produce would not escape. Second, the chances of his being able to write a self-replicating program that could survive mutation were close to zero. And third, whatever he did would surely take a very long time. Others, with greater technical experience, had been in the game longer, and had not succeeded. The gathering was extremely friendly, supportive, but not especially encouraging about immediate prospects. It was a reprise of Chris's review of the essay Tom sent five months earlier.

"I was aware of the security problem," Tom told me. "With the computer virus scare, we all were. I hadn't stressed it enough in my presentation. Chris and Doyne's advice was sound, that I must run my program on a virtual computer." What, I asked, is a virtual computer? "That's a real computer," said Tom, turning and pointing to one of the three machines on the table behind him. "And that's a real computer. So is that. A virtual computer is one that doesn't exist. You emulate it from software. You say, I'd like a computer with this set of features, and you write the features into the program, put in whatever parts of the computer you want for your simulation. The program effectively creates a computer inside a real one. That's why it's called a virtual computer." Like you can design a car on a computer, and test its performance? "The same kind of thing, yes." And if you have your creatures living in this virtual computer, there's no way they can escape? "That's right. You have to develop a new language for your virtual computer, and that was something new for me."

Tom willingly followed the Los Alamos group's advice on security, but was less influenced by their views on the difficulty of writing self-replicating programs that could survive mutation. The problem was known as "brittleness." In the exchange of E-mail letters earlier in the year, Chris had stated boldly that it "simply will not work." Any slight random change—a mutation—in the program would bring it down, he said. With characteristic stubbornness, Tom replied: "I'm not willing to ignore the approach just because nobody else has gotten it to work." Why, I asked, had they been so concerned? "Because no one had done it, I suppose," replied Tom, "and because Chris comes from the University of Michigan, where they'd had a lot of experience trying to make it work." And why were you so sure it would work? Tom shrugged, and said, "I just thought it might." Then he added: "Genomes evolve, so why shouldn't programs?"

With the last of the three concerns—that whatever he did would take a very long time—Tom fully agreed.

Tom returned from Los Alamos, handed in his dossier to the tenure review committee, and walked away from sixteen years as a field ecologist. He was going to play God, create a life in a computer, and become a naturalist of digital organisms. Or he would fail, and that would be the end of Tom Ray, university professor. This was mid-October 1989.

The task was to produce a simple organism that contained instructions for its own replication, and no more. Nothing about its potential evolution would be built in. The organism would be subject to a low rate of mutation—a flip from 1 to 0, or vice versa, in its code, just as Earth organisms experience random changes in their DNA. The organisms would compete for space and time: space in the computer's memory, an analogue for space in a real ecosystem; and for the amount of time the replicating algorithm would spend in the computer's central processing unit, an analogue for energy. "I wanted to avoid building anything into the system that might shape its behavior, that would determine the patterns of its behavior," said Tom. "I wanted it to have the simplest of constraints, variation and selection, the basis of natural selection." In the context of dynamical systems, I asked if he wanted to see what global patterns would emerge from the operation of local rules, the variation and selection. "That's precisely right."

Tom had already designed his simple organism—an eighty-instruction algorithm—before he went to Los Alamos. It had been "a trivial task," as the Go player of a decade earlier had suggested. The challenge next was to ensure that the whole thing didn't fail because of the brittleness problem, that any small mutation would bring the program to a halt. Inspired by further analogies from biology, Tom modified his computer system. First, he reduced the size of the instruction set of the machine code from something like

4 billion to just thirty-two. This brought it in line with the twenty amino acids (coded for by sixty-four codons in the DNA) that operate in the biological realm. "I just had the feeling that staying with the original huge number of code instructions would be a problem," he explained.

The second analogue Tom borrowed from biology was "addressing by template." In most machine codes, when a piece of data is addressed, the exact numeric address of the data is specified. This is not how biology goes about things. For example, a protein, A, in a cell will interact with a second protein, B, when the two come together by diffusion; complementary shapes on their surfaces lock into each other. Tom exploited this trick of nature, by putting a short code of four instructions, in the pattern 1111, at the head of his creature, and another group of four, in the pattern 1110, at its tail. "Between these two instructions I filled in a program that would start by looking for the pattern complementary to 0000 to find its head and record its location, then look for the pattern complementary to 0001 to find its tail and record its location; and then calculate the size," Tom explained. The program in between the head and tail codes contains instructions for replicating the organism and finding a nearby location for the "daughter" organism. Moreover, addressing by template also allowed organisms to find neighbors, with which they might interact.

As far as I knew, no one else had taken this path, of marrying tricks of molecular biology with tricks of computers, with the aim of producing artificial life. "I think it's important," Tom said. "I'm a pretty good programmer, for a biologist, but not compared with those guys at Los Alamos. But I know about biology; they don't." Tom had expected to spend years modifying the program. Instead, by 18 December, just two months after starting on it, Tom was able to send an E-mail message to Chris, saying, "My [Artificial Life] simulator is running!" He also told Chris that he'd decided to call the system Tierra, which is Spanish for Earth, rather than Gaia. "I didn't want to confuse what I was doing with all that New Age stuff," Tom explained to me.

Two weeks later, the last bugs were out of the system and it was ready to go. It was 3 January. "I set the thing going, and left it to run overnight," Tom said, recalling what obviously had been a tense but exquisite moment in his life. "I didn't sleep much." Tom had already glimpsed fragments of life in Tierra during the debugging process. He knew that something was going to happen, something interesting. But he had no way of predicting just how interesting it would be. "All hell broke loose," was how he described what had occurred overnight in his virtual world. "From the original ancestor, parasites very quickly evolved, then creatures that were immune to the parasites," said Tom. "Some of the descendants were smaller than the ancestral organism, some were bigger. There were hyperparasites, social creatures. I saw arms races, cheaters, there was—" Wait a minute, I interrupted, you have to explain these creatures to me. When Tom described himself as a naturalist of a virtual world, he meant it: the digital organisms were as real to him as the ant butterflies had been.

"OK," said Tom. "Would you like to see some of it?" In that first burst of evolution, Tom had to delve into the database to uncover the bestiary. Now, however, with the help of computer enthusiasts at Delaware, he had a visual display of Tierra. The different creatures are represented by horizontal bars of different lengths and colors stacked on the screen. Though no Walt Disney animation, this multicolored matrix nevertheless conveyed the sense of a world in motion, as new creatures entered the scene while others dropped out. "Let's look at parasite-host interaction," said Tom, as he clicked through a directory. The records of that first run are stored, and Tom can go over what happened again and again, just like a paleontologist searching through the fossil record of life.

Parasites, Tom explained, evolved by dropping a chunk of the original eighty-byte genome, finishing up just forty-five instructions in length, and making use of neighbors' replication instructions. They don't harm their hosts, but they deprive them of valuable energy and space. When hosts are plentiful and space is in short supply, the parasites flourish. A crash in host population is followed by a crash in parasites, too, just as in real life. "You get the classic Lotka-Volterra cycle," said Tom as we watched the periodic rise and fall of host populations, tracked closely by parasite populations. Textbook, I said. The cycle, which is the best known pattern in population biology, describes the interaction between populations of a predator species and its prey. With an established prey population, a population of predators will increase. Eventually, the predators begin to have a serious impact on the prey population, which begins to decrease. With fewer prey to eat, the predators begin to suffer, and its population decreases. Released from the pressure of predation the prey population now rebounds, followed by the predator population. The cycle of rise and fall of populations continues indefinitely, and the pattern was to be seen in Tom's digital ecosystem. "Yes, there's lots of textbook ecology in Tierra," Tom said. Competitive exclusion, keystone predator phenomena, periods of stability punctuated by bursts of change—many occur in Tierran ecology, all classic patterns of Earth ecology. "We even see occasional mass extinctions."

And all this emerges from a few fundamental rules, I ventured, nothing built in that would ensure these patterns? "Nothing built in," replied Tom. "What you're seeing is the emergence of global patterns from simple rules. The notion of something deep as an organizing force appeals to me, always has." That's a familiar sentiment, I said. Stu Kauffman used the same words when he described his Boolean networks and the emergence of order. "We've talked a few times," said Tom. "Nothing philosophical, though, just about the details of the systems, his and mine. But, yes, from what you say we both have the same sense of something deep here. That's why evolution has been a central scientific theme for me, the idea that some process on the level of physics leads to increasing complexity. That's what you see in nature, and that's what you see in Tierra."

The explosive evolution in Tierra that January day of 1990 took Tom by surprise, but it seemed a wonderful opportunity to demonstrate to the artificial life community what could be done with this unique blend of biological and computing principles. Just a month later the second artificial life conference would be held, this time in Santa Fe. Pursuing his obsession to establish the field as a legitimate scientific enterprise, Chris Langton had devoted much of his time since the first workshop to setting up the second. The press of people wanting to take part in the event, to show off their creations, was enormous, and Chris was constantly manipulating the program, cutting down the number of talks any individual might give, and trimming down the time for presentations. Tom had originally been allocated two slots, each of forty minutes. Ultimately he had just one, of twenty minutes. So, as time passed and Tom had more and more to tell he found he had less and less time in which to tell it. Still, he was going to wow them with Tierra, no question about that.

For Chris, the clamor to attend AL2, as it was called, seemed a vindication of his obsession. Proving that artificial life was more than a fantasy had been his goal even before he went to the University of Michigan in 1982. His thesis, on the dynamics of cellular automata, in reality had been something of a Trojan horse for his study of artificial life. And when he left in 1986, to join the nonlinear dynamics groups at Los Alamos at Doyne Farmer's invitation, Chris still hadn't completed the formal part of his graduate course, that of writing up the thesis. The romance with artificial life had been too distracting. The continuing romance, and the exigencies of organizing the two workshops, proved equally distracting at Los Alamos. As a result Doyne came in for a lot of pressure from the lab's bureaucracy over his protégé's inability to finish his doctorate. "Doyne was my protector," Chris told me. "I've always been lucky, with people protecting me, letting me do what I had to." (The thesis was eventually turned in and approved in 1991.)

Where did this obsession with artificial life come from? I asked Chris. "I can trace it back to a specific event, a bizarre experience," he began. In the early 1970s Chris was working in the Psychiatry and Psychology Research Laboratory at Massachusetts General Hospital, Boston, in his conscientious objector status from the Vietnam War. The lab had needed someone who knew about computers, so Chris jumped at the chance: he didn't know much about computers at the time, but the opportunity seemed better than pushing bodies about in the morgue, which was his initial role at the hospital. There was great camaraderie and tremendous intensity in the lab, and people often worked late into the night.

"One night I was there alone, really late, about three in the morning actually," said Chris. "I was sitting at my desk, debugging code, going over it with paper and pencil, trying to figure out why it wasn't working. I had the Game of Life running on the screen, and occasionally would look up and watch for a while, set it going again when it stopped. We all did that. It was new then." Conway's Game of Life had come out in 1970, and had fascinated everybody with its autonomy, its ability to produce complex patterns, its uncanny sense of having a mind of its own.

"Suddenly, I got the sense that I wasn't alone," said Chris. "A completely visceral feeling, hairs standing up on the back of my neck. I swivelled round, but no one was there. I thought maybe one of the monkeys had escaped from the cages. No. So I went back to my desk, sat down, saw that the Game of Life had petered out, and so started it up again. I suddenly realized that something on the screen must have triggered that feeling."

A cipher in your peripheral vision, I said. "Yes, it must have been," said Chris. "I let my mind follow the thoughts, and they had the feeling of being mysterious, not quite forbidden, but unexplored and dangerous." It was as if an idea had slipped surreptitiously into his brain and had begun to proliferate, spawning meta-ideas in all directions, unconstrained, adventurous. "I was staring out over the Charles River, toward Cambridge . . . car lights moving silently by the river . . . buildings with stark shadows from street lights . . . steam coming out of smoke stacks . . . a sense of all these behaviors out there, of the city alive . . . not people, not biology, just life." Chris paused, recalling the power of the moment.

"It was like a drug experience, when you run with a crazy fantasy," Chris began again. "You take down the usual mental barriers to crazy thoughts, and just let them develop freely. It was like a hurricane of ideas sweeping across the mental landscape, and I was just a spectator." Chris likened it also to a state that occasionally happens when he's playing guitar, when music takes fire and runs as if on its own. "I don't know how long it lasted, maybe two minutes, maybe two hours. But it went very deep. I got caught up with the idea of information having a life of its own, a living logic. It's irrelevant whether you'd say it's alive, but it's a similar class of phenomena." The hurricane passed, and Chris's mental landscape was irrevocably altered. He knew one day he would make artificial life a reality. The year was 1971.

"It's true that organizing these workshops has taken a lot of time out of my own research," Chris told me. "But getting the discipline going and respected, that's what matters. I don't care who does it, as long as it gets done. AL2 took us a long way toward that goal." I asked how much of an impact Tierra made. Did it wow everybody, as Tom had hoped? "You know, it didn't," said Chris, as we talked in his office at the Santa Fe Institute. "Maybe it was my fault, because I couldn't give him more time for his talk. But he'd only just got the results, and was still working on them while he was here, so maybe he hadn't polished his presentation. In any case the full implications didn't come across. But look at this," said Chris, as he turned and pulled down the conference volume. "Look at this work by Kristen Lindgren, a very different system, but one in which you get competition and selection." Chris had found Lindgren's article, and was leafing through it for a graph he wanted to show. "Look familiar?" he asked.

Lindgren's was the simplest of evolutionary systems, based on a famous game, the Prisoner's Dilemma. In the classic Prisoner's Dilemma, two players, arrested for a crime they both committed, are separated and offered a choice by the police: inform on your partner and receive a reduced sentence, or remain silent. If both remain silent, both go free, but if one partner informs, the other receives the maximum sentence. Games theoreticians have demonstrated that, even though the highest-rank alternative is always freedom through silence, informing, so as to reduce the risk of the maximum sentence, is the optimum strategy. In Lindgren's version, the prisoners play the game not once but repeatedly, with the possibility of making different decisions at each iteration. Lindgren allowed strategies to evolve (by a kind of mutation), often becoming quite complicated, as rounds of play passed. The different payoffs of these strategies are analogous to different fitnesses in biological systems and the different fitnesses of organisms in Tom Ray's system. The effect is competition among a population of evolving strategies, with the emergence of coevolving populations as in Tierra, but much more abstract. And a graph of the history of populations of strategies through time looks uncannily similar to the history of Tom's much more complex community of digital creatures.

Yes, I responded, it looks very familiar; it looks like Tom's data. "A population remains in balance for a while, then, wham, rapid change, you get chaos for a while, then more stasis," said Chris, describing the graph in front of us. "You even get mass extinctions," he said, "look." Sure enough, the populations sometimes took a nosedive. When I had visited Tom in Delaware, I'd asked him about the mass extinction events in his system, and said how very much it all looked like the history of life on Earth. "Yes, and no asteroids," he had replied emphatically. No asteroids, just the dynamics of a complex adaptive system. Can it be a coincidence, I asked Chris, that you see this sort of pattern in a simple system like Lindgren's, in a more overtly biological system like Tierra, and the history of life in the real world? "I don't think so," said Chris, cautiously. "I think what we're seeing is something deep, some fundamental dynamics of similar systems."

One of the patterns Chris described, of periods of stability interrupted by bursts of change, is well known to ecologists and, in more recent years, to evolutionary biologists, too. Punctuated equilibrium is the term used by evolutionary biologists. Stephen Jay Gould and Niles Eldredge, of the American Museum of Natural History, proposed the idea of punctuated equilibrium in 1972, and provoked an at-times acrimonious debate as a result. Two decades later, there are still some who doubt its reality, but most accept that it is at least part of the overall pattern of evolutionary history.

"The pattern of punctuated equilibrium always reminds me of flow of liquid through a pipe," said Chris. "At low velocity you have smooth flow. At high velocity you get turbulence, chaos. Just as you switch from turbulence to chaos you have a period when the flow is smooth, then this cell of turbulence comes along; then smooth flow resumes for a while; then more turbulence. It's called intermittency." Is intermittency like the edge of chaos? I asked. "It's a reasonable analogy, perhaps more than an analogy." Does that mean that a pattern of punctuated equilibrium in an ecosystem or in an evolutionary history implies that the systems are at the edge of chaos? "I think it might." Not definitive? "Not definitive."

After his disappointing showing at the artificial life workshop, Tom returned to Delaware, delved even deeper into the analysis of life in Tierra, and prepared for what turned out to be a year of extensive travel in the United States and Europe. He took his virtual world with him, giving seminars in university departments. In November alone he presented Tierra in Aarhus, Copenhagen, Basel, Montpellier, Paris, Nottingham, Oxford, Cambridge, and Sussex. "I'd usually begin saying that 'We're all biologists; we're interested in evolution, but we only have one example to study, the one of which we are a part,' " Tom told me. "Then I'd say, 'Well now we have an opportunity to explore other worlds, other examples of evolution.' At this point a few people usually started to snicker. Then I'd show them Tierra, and most if not all of the audience would be hooked."

Biologists are often skeptical of mathematical models, suspicious that simplification might bring comprehension at the expense of reality. With a model that its author promised would give a glimpse of the processes underlying the entire history of life, audiences should feel doubly justified in being skeptical.

"I thought it was wonderful," said Richard Dawkins, when we met in his college rooms in Oxford. In November 1990 Tom had given a seminar in the university's Department of Zoology, perhaps the highest concentration of top evolutionary biologists in the world. "Sometimes it takes a while for the full import of Tom's work to sink in, but I was prepared and immediately realized how important this was." Five years ago Richard and an astrophysicist friend had schemed how they might produce a self-replicating, mutating, adapting world, like Tierra. "It was uncanny how close our ideas were to what Tom actually produced," said Richard. Why didn't you pursue it? I asked. "It seemed a very big project, a very ambitious bit of programming. I think it seemed too big a job."

In his 1986 book, *The Blind Watchmaker*, Richard described a system he developed instead, a program that produced patterns from simple rules. The patterns, which he called biomorphs, evolve, but only through artificial selection: the computer generates mutants from a parental form, but the computer operator must choose among the variants which one goes on to the next stage of mutation. Extremely lifelike patterns emerge, hence their name, but, unlike Tom's system, without human intervention they go nowhere. "Until we master interstellar travel, Tom's system, or something like it, is the best chance we have of studying another example of evolution," said Richard. "He's created a silicon universe."

Another example of evolution? Is that really the aim of Tierra? I asked Tom. "Yes it is. We've got countless products from the one example of evolution we know, the one based on DNA, and we can learn a lot from that. But we'd like to know how general it is, because that would tell us something about the organizing principles of evolution." I'd read somewhere that Tom's ambition is to rerun the Cambrian explosion. I asked him why. "Isn't it every evolutionary biologist's?" he replied. "Most people, if asked, would say that the most important event in the history of life on Earth is its origins, and of course that's true in a way. But I'd argue that the Cambrian explosion is an event of equal importance. It's where all the interesting biology begins, the interesting evolutionary patterns." That first run of Tierra was like a Cambrian explosion, wasn't it? "In a way, but what we need to put in there are multicellular organisms, to see cellular differentiation and the emergence of morphological complexity. Then we might be able to see if there's an infinity of possible worlds or perhaps just a few." I asked if he would be able to produce a Cambrian explosion before I finished my book. " 'Fraid not. It's a big task."

Even so, I said, life on Tierra has sufficient similarities to life on Earth to encourage the hope that they share some fundamental properties. Particularly the mass extinctions. If mass extinctions can come about in a system in the absence of asteroid impact and without wiring in assumptions about connectedness among species, then this surely is a significant observation. I asked Tom if he thought all mass extinctions might be the result of the dynamics of a complex adaptive system. "No, I don't," he said. "The evidence for at least some impact-induced extinctions looks pretty convincing to me. But if Tierra is telling us anything, it's telling us that the dynamics of complex systems can produce patterns we would not have predicted, patterns that we see in nature, and that includes extinctions of significant size."

In the early fall of 1990 Tom called Stu Kauffman, and invited him to give a seminar at Delaware. (Tom's ulterior motive was that he wanted to spend some time at the Santa Fe Institute, and had heard that Stu was influential there.) Before the seminar Tom showed Stu Tierra. "I thought it was wonderful stuff," Stu told me. "Tom has brought an ecologist's view to complex systems, and we needed that. The buildup of diversity he sees, it's a neat story of compounding complexity." Naturally, the edge of chaos concept came up. "If there's anything to the concept, then we should see some evidence in Tom's system," said Stu. What would you look for? I asked. "What do you think?" A power law distribution? "That's right. A power law distribution of extinctions."

Stu suggested to Tom he might take a look at the size and frequency distributions of extinctions in Tierra. Tom, somewhat skeptical of "fashionable" notions like the edge of chaos and downright suspicious of the significance of power law distributions, didn't plot the data right away. About nine months later, just before visiting the institute, he did the plots, and handed them to Stu when he arrived. "I was stunned," said Stu. "He'd plotted thirty thousand extinction events, and look, this is what he got." Stu pulled out a sheet of paper showing a slightly convex curve. That's like the curve you got when you plotted Dave Raup's extinction data, I said. "That's right, a power law, slightly curved, suggesting that the system is just into the frozen regime, near the edge of chaos."

Stu was particularly impressed with the result because with his own coevolutionary system he'd had to construct an explicit model of fitness landscapes and wire in which species interacted with which other species. "True, with my model I know why the system goes to the edge of chaos: that's where fitness is optimized," said Stu. "My guess is that Tom's critters have got themselves to the edge of chaos, and for the same reason." I asked if he could prove it. "Not yet, but we're cooking up ways to try."

Tom, now much less skeptical than he was of the concept, thinks it might be possible to tune his system to the edge of chaos by altering the mutation rate. "Mutation rate in my system is somewhat analogous to the lambda parameter [the mathematical device that sets the rules of the cellular automaton and allows the consequences to be monitored across a continuum] Chris used in his cellular automata," Tom told me. "If I turn up the mutation rate, the system should go chaotic and die out. At a low rate nothing very interesting should happen. In between these two rates we should see a rich ecology produced, and if this is the edge of chaos, this is where we should see avalanches of extinctions with a power law distribution." You haven't done that yet? "Not yet," said Tom. "Sorry." But the system you already have displays that kind of power law, I said. Doesn't that mean that Tierra might have evolved to the edge of chaos all by itself? "Yes, it's possible, but I'd like to run a test to make sure."

Was it mere coincidence, I asked Stu, that we see a power law distribution (or something very close) in extinctions in the real world, in your coevolutionary model, and in Tierra? Or were we seeing a common signature of the same fundamental processes? "Look," he said. "We're in unknown territory here, the whole science of Complexity is in a way. We're building a case, bit by bit. I think that coincidence—call it what you will—is part of the case. Don't you?"

Well, I said, it looks suggestive. But I need to look at real ecosystems first.

(Yes, Tom did get tenure.)

CHAPTER SIX

Stability and the Reality of Gaia

"Hedgerows are a thousand years old hereabouts," said Bill, as we hurtled at a hazardous pace along dark country lanes, deep in the English countryside. "Two thousand years, some of them." With alarmingly small clearance on either side of the car, the ancient hedgerows, anchored in a base of earth and stone, towered a good six feet above us. The headlights scooped a fleeting moment in the seemingly endless winding trench as we sped onward. I should not have been anxious, as Bill has done this journey many times, ferrying visitors from the railway station in the Devonshire town of Exeter to the tiny village of St. Giles on the Heath. Bill is the local taxicab company in St. Giles, and represents the only form of transport in these parts, some eighty miles from the southwesternmost tip of the British Isles.

The journey took almost an hour, beginning in the fading light of what had been a bright February day, and finishing in the kind of darkness experienced only in deep countryside. "That's Dartmoor over there," Bill said, half an hour into our journey, indicating the rising, featureless terrain to our left. It was dusk by this time, perfect for glimpsing so forbidding a place, my imagination fed by Conan Doyle's *Hound of the Baskervilles* and other horror stories set on the moor. "Someone got themselves lost there a while back," observed Bill. "Still no sign of them."

I struggled to capture the words in Bill's rich Devon accent, rounded and rapidly spoken. "The wife and me, we have picnics there in the summer," he added incongruously, as if part of the same thought.

We had passed St. Giles on the Heath a mile back, and were now negotiating ever-narrower lanes, grass growing down the center. "Not long to go," Bill assured me. Then, as we crossed a cattle grid, the headlights caught the sign, "Coombe Mill Experimental Station." A "hazard" notice of the sort you see in research laboratories hung on a five-bar gate. And as we stopped in front of the old millhouse the headlights came to rest on a white marble statue, the figure of a woman. "Yes," said Bill as he saw me looking. "That's Gaia."

The door of the cottage opened, and a man stepped forward to greet me, hand outstretched: "Hello, I'm Jim Lovelock," he said, with a soft voice and a gentle, almost diffident smile. In his early seventies, white-haired, he projected a combination of vigor and extreme courteousness. Was this the man whom much of the biological community regard as the devil incarnate, a threat to the integrity of true science? We were joined by Sandy, his wife. "Come inside," she said. "I'll get us all a drink."

Over dinner in a low-beamed dining room Jim talked of his early days in science, when as a postdoctoral fellow at Harvard he had to sell his blood once a month to supplement his meager stipend. "Fortunately I have a rare type, so I got fifty dollars each time." He found Harvard bureaucratic, rigid, and exploitative. "When the fellowship was up they asked me to stay another year," Jim recounted. "I refused, so they offered to double my salary. I still refused. They'd triple it, they said, quadruple it. They hadn't thought to give me more in my first year when I so obviously needed it, so I said, 'I'm off.' "

The phrase "I'm off" captures much of Jim's free-spirited approach to life and work: after two decades at the National Institute for Medical Research in London, in 1964 he turned his back on the conventional research establishment, and set himself up as an independent scientist, first in a thatched cottage in Wiltshire and now at Coombe Mill. "I'm an inventor," he explained. "My science is intuitive. I can be more creative in this kind of setting." Through the years he supported his family and his research by developing and patenting about thirty analytical and control devices. The first of them, the Electron Capture Detector, remains among the most exquisitely sensitive means of measuring atmospheric chemicals, including significant pollutants such as chlorofluorocarbons, or CFCs.

After dinner Jim directed me to the guest house (a converted barn), whose occupant a few weeks earlier had been Hugh Montefiore, former bishop of Birmingham. I felt in good company. "We'll meet early in the morning," said Jim. "We have a lot to talk about."

When, two years ago, I embarked on my exploration of the relevance of the new science of Complexity to the patterns of nature, I had not guessed it would bring me to Jim Lovelock, inventor of the Gaia hypothesis. I did know that I would be led through the complexities of embryological development and evolution, and into the dynamics of extinction. I suspected I would discover its imprint in the operation of ecosystems. And I speculated that complex societies—the rise and fall of civilizations—might also be driven by the engine of complexity. But the entire globe? The intimate interaction between the biological and physical worlds that, according to Gaia, pulse as a single organism? I should have foreseen it. So, when Stu Kauffman a year earlier described the drive to the edge of chaos in his coevolutionary models as "mini-Gaia," it provoked one of those "Of course!" experiences. Why hadn't I thought of it earlier?

If much of nature dances to the tune of complex dynamical systems, then the consequences should be apparent from single organisms through to the way the entire planet works. The phenomena of spontaneous generation of order and of adaptation to the edge of chaos would shape what we see, level built upon level, a hierarchy of effects, with Gaia as its ultimate expression. If true, that is.

The evidence I had seen to this point was strong enough to encourage me to go further, to take the step toward Gaia. "But that's flaky nonsense," some of my biologist friends remarked with a mixture of amusement and concern that I might be losing my grip on reality. I tried to explain my rationale, the explanatory power of complex adaptive systems, and the logic of including Gaia in such a scheme. I realized I was straying far from the territory of conventional biological wisdom, and was neither surprised not discouraged by their blank looks. "Well, at least have fun," they said.

Before I set off for England I asked Stu to elaborate on how he thought his coevolutionary model reflects on Gaia. "You have to understand that I'm no expert on Gaia," he said. "But as I see it, Lovelock is arguing that Earth's biological and physical systems are tightly coupled in a giant homeostatic system. My coevolutionary model is a clue that coevolving entities such as he talks about can control the structure of their landscapes and how richly coupled they are." You mean fitness landscapes, not real physical landscapes? "Yes, fitness landscapes," replied Stu. "You breathe out carbon dioxide as garbage, and that plant behind you breathes out oxygen as garbage. Where did that functional integration come from? And on a global scale it all balances. Isn't that remarkable."

But, I asked, is it more than mere analogy to talk about Gaia in the language of complex systems, to think of a self-organized, giant self-regulating entity? "It's not unreasonable to think there might be an attractor to the meta-dynamics of the system," responded Stu. "The adaptive agents collectively make the worlds they live in congenial to themselves, and are drawn to that characteristic structure, the edge of chaos, where their interests are mutually balanced. Now *that's* homeostasis."

OK, I said, how do I discover whether Gaia lives, or is just a figment of Jim Lovelock's imagination? "You need to know whether systems are coupled together," Stu replied after a few moments' reflection. "You need to know how extensive the links are, because if they're short, you won't have a global system. You need to get a sense of a dynamical system that has emergent properties, properties that might lead to global homeostatic mechanisms." Any more? "Yes, you also need to talk to a good ecologist, someone who knows about ecological communities, not just predator-prey pairs chasing each other around."

Coombe Mill is a harmonious mix of eighteenth-century charm and twentieth-century high tech. Jim was in the twentieth century when I arrived at the cottage after breakfast, early as arranged. "Come in. Let me show you something." The room was full of computing equipment, with barely enough space remaining for two people to sit and talk. A converted cowshed fifty yards from the house serves as a workshop, where Jim builds his analytic inventions. This room is for inventions of another sort.

"Look," Jim said, indicating a computer screen. "If I have a world devoid of life, and the solar luminosity increases, the global temperature goes up steadily." A line climbing steadily at about forty-five degrees showed the temperature rise. "Now watch what happens if I put some seeds in here, for white daisies and black daisies." Another set of curves came up on the screen. I could see the black daisies begin to proliferate while the solar luminosity was still quite low, and then begin to decline as the sun pumped in more and more heat. As the growth of the black daisies fell off, white daisies began to multiply. "Now look at the global temperature," said Jim. "Isn't that interesting?" Instead of an inexorable climb as the model's sun heated the world, the temperature graph looked like a step: up, level, then up again. "Very early on, when solar luminosity is still quite low, the temperature is brought up to about twenty-three degrees Celsius, which is the optimum for growth of the daisies, and it remains more or less at that level for a long time, until suddenly it shoots up," Jim explained. "That's Daisyworld."

Just having those two types of daisies responding to light and heat from the sun maintains a level temperature? "It's a very simple model, of course, but it has a very powerful message,"

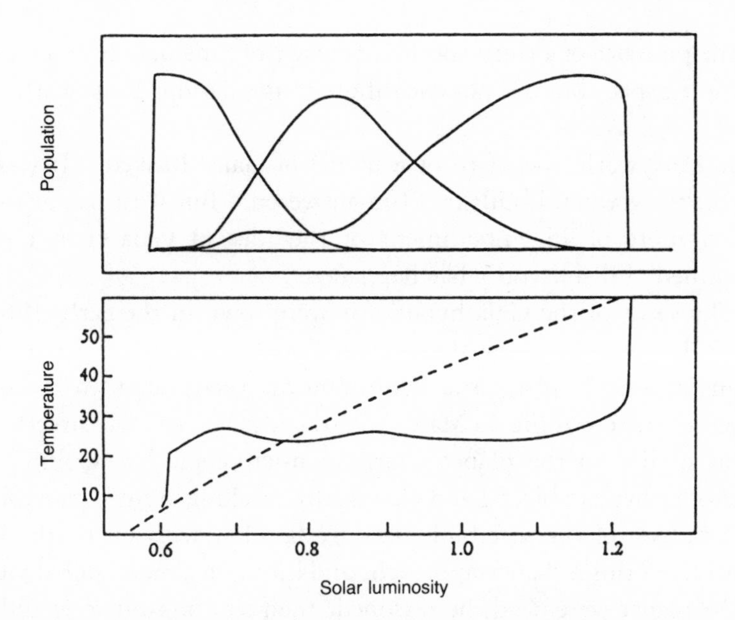

Fig. 6. Modified Daisyworld: the top panel shows changes in the populations of dark daisies (left), white daisies (right), and intermediate-colored daisies (middle) as luminosity increases (shown on horizontal axis). In the bottom panel, the upward sloping, 45-degree line shows the increase in global temperature that occurs in the absence of daisies; the S-shaped curve shows the global temperature under the influence of a Daisyworld—it remains relatively level and close to the 22.5 degrees Celsius that is optimum for daisies. Courtesy of James Lovelock.

Jim replied. "You see, it's a population biology model in which the different-colored daisies are competing for space in which to grow. The black daisies have an advantage when sunlight is feeble, because they can trap the heat and warm the planet. But too high a temperature suppresses growth, and then the white daisies have an advantage because they reflect light; they increase the albedo of the planet. The result is that the temperature is maintained close to the optimum for growth of the daisies, until the sun gets too hot and the whole thing collapses." That's like homeostasis, isn't it? "It *is* homeostasis," Jim replied. "And it's an emergent property of the system." Jim went on to show me that the same effect occurred even in the presence of a third species of daisy (intermediate in color) that occupied space but did not contribute to regulation. Daisyworld, he said, is robust.

Is Daisyworld meant to be a model of Gaia? I asked. "It wasn't meant to be when I built it," Jim answered. "But it turns out to be much more of an embodiment of the idea of Gaia than I ever imagined. I'll tell you what happened."

The seeds of the Gaia hypothesis were sown in the early 1960s, when Jim was hired by the National Aeronautical and Space Administration (NASA) as a consultant in their quest to discover whether there was life on Mars. NASA's idea was to look directly for signs of life on the planet's surface: microscopically, looking for microbe-shaped objects; and chemically, seeking signs of microbial metabolism of the sort biologists are familiar with on Earth. Jim considered this a chancy approach, and hit upon a more global view. If the planet were dead, he reasoned, then its atmosphere would be determined by physics and chemistry alone; it would be in equilibrium with the chemistry of the planet's minerals. But if life lurked there, however simple, it would undoubtedly exploit the atmosphere for raw materials, thus changing its chemical composition. A living planet would have an atmosphere shifted away from a simple equilibrium with chemistry and physics of rocks. Simple argument; compelling strategy; ignored. NASA chose the chemical analysis route, and when the Viking lander sent back results in 1975, they were ambiguous at best.

"There's no life on Mars," Jim told me. "I knew that from the spectral analysis of the atmosphere. I tried to tell them, but they weren't interested in hearing what I had to say."

The task of trying to figure out the characteristics of a living planet from afar established in Jim a "top down" way of thinking about Earth and its dynamics. "For many people, the image of the Earth as seen from spacecraft—a dappled blue-and-white sphere—was an emotional and germinal experience, a first glimpse of the planet as a whole," he said. "I'd already come to that through thinking about atmospheric gases and what they indicate about the activity of the planet." Then, one afternoon in 1965, Jim experienced one of those intuitive leaps that for him is the stuff of science. "I knew that the composition of Earth's atmosphere had remained stable over long periods of time. I also knew that there was a continuous turnover of gases, of oxygen and carbon dioxide in particular. What, I wondered, controlled the long-term stability?" The fact that the Sun's output has increased by 25 percent during the history of life on Earth made atmospheric stability yet more of a puzzle.

"My intuition was that life provided the controlling hand," said Jim, "in an active partnership with the physical world, controlling atmospheric composition and global temperature." Like in Daisyworld? I said. "Yes, like Daisyworld, but on a much bigger and more complex scale, of course." Jim's principal scientific preoccupation then, as now, was in atmospheric chemistry and invention of analytical tools, for which work he has been awarded England's highest scientific honor, fellowship of the Royal Society of London. The notion of global atmospheric control—of physical and biological worlds in a close mutualism—had, however, insinuated itself deep into his mind, a preoccupation for which he is viewed with profound suspicion by his fellow scientists. "One day in 1969 I was working outside my cottage and William Golding came by, just setting off for a walk. I asked if I might join him." William Golding, the novelist? "Yes, he lived nearby. This was in the village of Bowerchalke, in Wiltshire. During the walk he asked what I was doing, and I told him my ideas about atmospheric homeostasis. Golding had been a physicist; not many people are aware of that. Anyway, he said, 'For so grand an idea you need a grand name. You must call it Gaia.' "

For the next half hour a grand misunderstanding unfolded, as Golding had in mind the Greek goddess Gaia, Earth Mother, while Jim thought he had said Gyre. "Gyre are great eddies in the ocean, self-organized, big, and controlling, and that seemed reasonable," Jim told me. "Finally it became obvious that we were talking about different things, and Gaia it became." Didn't you think that naming a serious hypothesis after a Greek goddess might be a problem to your scientific colleagues? "I didn't," Jim admitted. "It seemed such a powerful idea."

But it was a problem. Not only did the hypothesis extend beyond the boundaries of any single discipline—always an obstacle to comprehension in the tightly compartmentalized world of science—but it also seemed to imply teleology, a sense of purpose embodied in the whole system. In 1972, for instance, now with Boston University biologist Lynn Margulis as ally, Jim stated the Gaia hypothesis thus: "Life, or the biosphere, regulates or maintains the climate and the atmospheric composition at an optimum for itself." The phrase "for itself" flagged the idea as teleological, as implying purpose. As a result, most papers on Gaia could not break into the conventional scientific press. The fact that Jim sought other means by which to promulgate his idea—namely through articles in popular science magazines and books—served to convince most scientists that Gaia was indeed unscientific.

Jim admits that some of his popular writing was a little "poetic." For instance, in his 1979 book, *Gaia: A New Look at Life on Earth*, he wrote that the coming of *Homo sapiens* had changed the nature of Gaia: "She is now through us awake and aware of herself. She has seen the reflection of her fair face through the eyes of astronauts and the television cameras of orbiting spacecraft. Our sensations of wonder and pleasure, our capacity for conscious thought and speculation, our restless curiosity and drive are hers to share." That *is* poetic, I said. "True," he replied reflectively. "I'm really a hard scientist, and this sounds like heresy." After a short pause, he said: "God damn it, when you get a good idea in science, it's pure intuition and that's often extremely difficult to describe. If I'd known then what I know now, I wouldn't have written it like that. But I'm glad it became controversial. The worst thing that could have happened was for people to ignore it."

They didn't. Many attacked it vigorously. "Pseudoscientific myth-making," was how British biologist John Postgate characterized it. Richard Dawkins argued that the hypothesis was fatally flawed, something that "would have instantly occurred to [Lovelock] if he had wondered about the level of natural selection process which would be required in order to produce the Earth's supposed adaptations." As there is but one planet Earth, Richard argued, there was no possibility of competition among Earth-like bodies, and therefore no possibility of natural selection forging the kind of homeostatic mechanisms that constitute Gaia. Period. Ford Doolittle, a geneticist at Dalhousie University, Canada, said in a review of Lovelock's 1979 book: "It is not novel to suggest that life has profoundly changed the Earth, but it is novel and daring to suggest that it has done so in a seemingly deliberately adaptive way, in order to ensure its own existence."

The criticisms cut deep, particularly the suggestion that Gaia was purposeful. "Neither Lynn Margulis nor I have ever proposed a teleological hypothesis," Jim said. "It's true that some of the things I've written have been imprecise, and this was eagerly interpreted as meaning purposefulness in Gaia. Doolittle's and Dawkins's criticisms really set me back. I was depressed about it for a year. I needed to be able to demonstrate to others what I knew intuitively about Gaia—that homeostasis emerged as a property of the system." As an inventor of control systems Jim has a deep intuition about them, but says repeatedly that they are often difficult to explain to others. In this case he had to invent something that would explicate the workings of a natural control system, the entire global system. He worried about it a lot. "Then, at Christmas 1981, it came to me, fully formed the way these things often do," he told me, clearly recalling the relief he felt at that moment. "It all seemed so obvious to me. I just sat down and wrote the program in an hour." Daisyworld? "Yes, Daisyworld."

But, I said, Daisyworld looks such a simple system. Does it really demonstrate the validity of Gaia? "Remember what I was trying to show. I said that the biological and physical worlds are tightly coupled, and that the biota operates in such a way as to ensure optimum physical conditions for itself. I had in mind a biological system that works according to conventional evolutionary rules, and that, like all complex systems in the universe, it has a tendency to produce stability and to survive. I needed to show that the stability emerges from the properties of the system, not from some purposeful guiding hand. Daisyworld does that."

I asked how he could be sure that more complex worlds would also be stable. "Wait a minute," Jim said, as he searched the computer's directory, finally finding what he wanted. "This one has twenty daisies, different shades between white and black. Tremendous stability." It looked very convincing. "I can be as complex as you like," said Jim, offering a challenge. What about different trophic levels, with herbivores and carnivores? "Would twenty species of daisies, five of rabbits, and three foxes suit you?" OK, I said. That would represent three trophic levels: primary productivity (the daisies), herbivores (the rabbits), and carnivores (the foxes).

That, I suggested, was quite a challenge for a population biology model, wasn't it? "Well, I have to admit that I had already generated the various forms of Daisyworld before I read much of the literature," Jim said with a chuckle. "I often do that, and it's lucky. If I had read the literature I would have discovered that working with models like this is virtually impossible with more than just a few species, because they go chaotic. Perhaps I wouldn't have tried if I'd 'known' it wouldn't work." But it did work. Again, the model biosphere—with daisies, rabbits, and foxes—interacted with the physical environment, and temperature regulation was the result. "Watch what happens if I disturb the system, by killing some of the daisies," said Jim. The daisy population dipped briefly, and the rabbit and fox populations followed, briefly. Blips occurred in the otherwise level temperature trace, too, briefly. "You see, the system can withstand disturbances," said Jim. "Stability is what I see in my systems, not chaos."

Why, I asked, did Daisyworld work like this, when all the best population biologists "know" that it can't? "Mostly, theoretical ecologists ignore the physical and chemical environment in their models, and that's a very important part of species worlds. Let me show you something." He pulled out a book by Alfred Lotka, *The Elements of Physical Biology*, published in 1925. Lotka is the father of

Chip Wills, University of New Mexico: "You don't get the feeling of a people scratching a living [at Chaco Canyon]. You sense an exuberance, a people capable of organizing tremendous feats of construction, including irrigation and farming under challenging circumstances."

V. Wills

© P. Crown

Patricia Crown, niversity of Arizona: By A.D. 200 pottery came important [for the Anasazi]; irrigation started, dentarity, too, more complex social organization. Something happened to oduce a big change. d it happened fast."

© California Institute of Technology

Murray Gell-Mann, California Institute of Technology: "In biological evolution experience of the pa is compressed in the genetic message encoded in DNA. In the case of human societies, the schema are institutions, customs, traditions, and myths. They are in effect, kinds of cultural DNA."

© R. Le

Pueblo Bonito, as seen from the northern rim of Chaco Canyon. The many round structures are kivas, ceremonial sites. Bonito was the largest of the Great Houses in the Chaco Canyon community; it was abandoned some time between A.D.1150 and 1200.

© R. Lewin

e brickwork in Chaco Canyon construction was meticulously assembled; a hallmark of the hitecture is the occasional T-shaped door (seen here at Pueblo Bonito) or window.

© R. Lewin

Roof beams were set into the wall structure, often in threes, as seen here at Pueblo Bonito. These beams, made from trees brought into the canyon from fifty miles distant, decay only slowly because of the arid climate. They have provided important material for tree-ring dating.

© Carrie D

Jeff Dean, University of Arizona: "The striking thing about Chacoan architecture is that the buildings pop up out of the ground, literally."

Warren McCulloch, Massachusetts Institute of Technology: In 1967 he told Stuart Kauffman it would be twenty years before anyone would take note of Kauffman's discovery of "order for free" in Boolean networks. He was right.

M.I.T.

© John Farnham

Brian Goodwin, Open University, England: "The creative principle of nergence is a deep mystery in many ways, it's true, and that's a property of complex dynamical systems. But timately it is intelligible. You can't say that about neo-Darwinism."

Stuart Kauffman, University of Pennsylvania: "If the new science of Complexity succeeds, it will broker a marriage between self-organization and selection. It'll be a physics of biology."

© University of Pennsylvania

© R. Lev

John Maynard Smith, University of Sussex, England: "You can't study nature without knowing there are bizarre adaptations out there, complicated ways of life that seem to fit an organism to its environment. So the problem becomes, how do I explain it? Adaptation by natural selection is the answer."

Chris Langton, Santa Fe Institute: "The edge of chaos is where information gets its foot in the door in the physical world, where it gets the upper hand over energy."

:ary Herz

© Brookhaven National Laboratory

Per Bak, Brookhaven National Laboratory: He says of the edge of chaos and self-organized criticality, "We're talking about the same kind of phenomenon."

© Patricia Evans

David Raup, University of Chicago: "As much as 60 percent of all extinctions may have been the result of asteroid impact."

© R. L

Tom Ray, University of Delaware: "If Tierra is telling us anything, it's telling us that the dynamics of complex systems can produce patterns we would not have predicted, patterns that we see in nature, and that includes extinctions of significant size."

James Lovelock: Mostly, theoretical ·ologists ignore the ysical and chemical ıvironment in their nodels, and that's a y important part of species' worlds."

© Sandy Lovelock

ırt Pimm, Univer- of Tennessee: "It's r to me that we e to think of :ies as being ›edded in complex amical systems, this gives you a · different view of world."

© S. Pimm

© University of Chi

Robert Richards, University of Chicago: "[Stephen Jay Gould's] rejection of the idea of prog
both for himself and for Darwin, is influenced by ideology."

©Bob Kalmbach. Courtesy of the University of Michigan News and Information Services.

Dan McShea, Univ
sity of Michigan: "
quickly learned tw
things. First, there
general though vag
consensus that
complexity has
increased through
evolutionary histor
Second, *complexity* i
very slippery word.
can mean many
things."

Michael Ruse, University of Guelph: "Scratch any evolutionary biologist and you find a progressionist underneath."

versity of Guelph

© Harvard University News Office

phen Jay Gould, Harvard
Jniversity: "Ours is a very
'brain-centric' view of
evolution, a bias that
orts our perception of the
true pattern of history."

© Car

Doyne Farmer *(left)* and Norman Packard, Prediction Company: Farmer was Chris Langton's "protector" at the Los Alamos National Laboratory; Packard: "People don't like [progress in evolution] for sociological, not scientific, reasons. I don't impute a value judgement to computational superiority."

Richard Dawkins, Oxford University: "I'm hostile to all sorts of mystical urges toward greater complexity."

Lloyd

© Susan Dennett

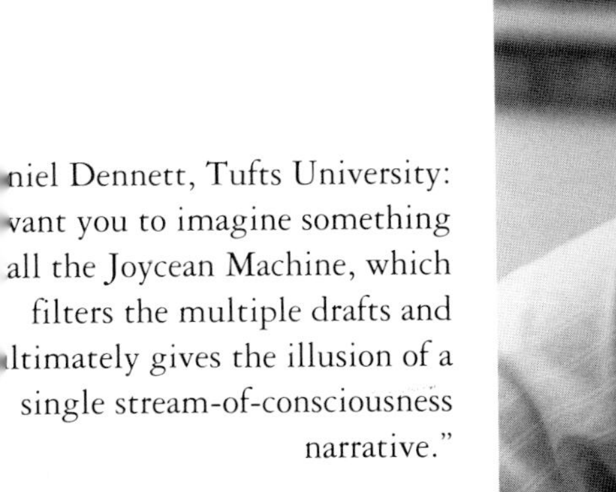

niel Dennett, Tufts University: vant you to imagine something all the Joycean Machine, which filters the multiple drafts and ltimately gives the illusion of a single stream-of-consciousness narrative."

Photo by Clint Clemens. ©Thinking Machines Corp.

Danny Hillis, Th
ing Machines Co
Hillis describes th
world's most adva
computer, the Co
nection Machine-
"trivial in comple
compared with th
brain of a fly."

© Becky

Patricia Churchland, University of California, San Diego: "When you think about brain activity it's correct to think about emergent properties at higher levels that depend on lower-level phenomena in the system."

Colin McGinn, Rutgers University: "The deep feeling of mystery we experience with respect to consciousness should at least encourage us to explore the possibility that an understanding of it is simply closed to us."

{c}Ginn

© Tim Wainwright

Nicholas Humphrey, Cambridge University: "It's the present that's crucial in consciousness, not reflecting on the past or the future."

© J. Cowan

Jack Cowan, University of Chicago: "There may well be some universal truths in the theory of Complexity, but the model still needs to be formulated with physics and biology in mind to do it properly. So far, that's lacking."

© Harvard Univ

Edward 0. Wilson, Harvard University: "It's time to look at the whole once again, and, yes, I think we can begin talking about insect colonies as superorganisms, but without the mysticism."

population biology, and the classic Lotka-Volterra cycle describes the periodic fluctuation of populations in simple predator-prey pairs. "Everybody knows about Lotka, but they seem to have forgotten this," said Jim. He directed my attention to a short passage:

> It is customary to discuss the 'evolution of a species of organisms.' As we proceed we shall see many reasons why we should constantly take in the view of evolution, as a whole, of the system (organism plus environment). It may appear at first sight as if this should prove a more complicated problem than the consideration of a part only of the system. But it will become apparent, as we proceed, that the physical laws governing evolution in all probability take on a simpler form when referred to the system as a whole than to any part thereof.

"Interesting?" inquired Jim. Very, I replied. "Lotka knew that the physical world was a vital part of the equation, but didn't have the computing equipment to do even the simplest Daisyworld, and no one else tried, until I did."

So, I asked, has Daisyworld done what you'd hoped, that is, persuade critics that Gaia does not include a purposeful motive? "Difficult to know," Jim said with a shrug. "I can't get it published in a scientific journal. I've tried *Nature* twice, but the reviewers were very dismissive. I don't think they were prepared to assimilate how much I have achieved, and they try to pretend it's worthless." (Jim is correct in this: I asked Robert May, a leading theoretical ecologist at Oxford University, his view of Daisyworld. "A marginal note on a more professional enterprise," he said. Richard Dawkins told me that Daisyworld "produces an illusion of control.") "So I had to publish Daisyworld in my second book," Jim continued. "I've given talks at conferences, and I find climatologists much more receptive to the whole concept. Climatologists are less reductionist than biologists, and more familiar with complex systems. That's why they have a better understanding."

Illusion or reality? I could see, as Jim had described it to me, that the Gaia hypothesis did satisfy some of the criteria of complex adaptive systems that Stu Kauffman had outlined. Specifically, the emergence of homeostatic mechanisms, which Stu had described as possibly a consequence of a system adapting to the edge of chaos. This, surely, is sufficient to persuade serious scientists to take the hypothesis seriously, I thought. Most important, however, was that Gaia should have some predictive power. Does it? I asked Jim. "You know what William James said about the fate of any new idea: 'First, it's absurd, then maybe, and last, we have known it all along.' For some things Gaia is in the second stage, for others it's in the third, so some of the predictions must be correct."

For instance, the hypothesis talked of long but strong links between tropical forests and climate: no rain, no trees, but equally, no trees, no rain. "And now it's hard to pick up a newspaper these days and not read about this kind of relationship, isn't it," said Jim. "The biological mechanisms for pumping down atmospheric levels of carbon dioxide, thus cooling the planet, are consistent with Gaia. And Gaia theory also led to the identification of possible global-climate control through the emission of dimethyl sulfide from the oceans. This may turn out to be as important as the greenhouse effects of carbon dioxide and methane."

Jim said Gaia *theory*, I noted, not Gaia *hypothesis*. In science the distinction is important. A hypothesis may be thought of as a loose framework of ideas, something to guide the direction of questions. When answers to the questions begin to support the hypothesis, the framework becomes strengthened, and eventually deserves the appellation *theory*. There's a theory of gravitation, for instance, and a theory of evolution. But a theory of Gaia? Is that what you meant? I asked. "That's right," Jim replied confidently. "With the observations made in the real world and the power of Daisyworld, I think Gaia deserves to be called a theory. Don't you?"

By now it was almost time for the obligatory midmorning walk. "I want to show you just one more thing before we go out," said Jim. He brought up a Daisyworld populated by a hundred species, but this time held the sunlight constant. "Watch what happens to the number of species." As time went on daisy species began to drop out of the global population, until the system settled at an equilibrium with just two. "Now I'm going to put in a step increase in solar radiation, 4 percent, which is equivalent to the change between glacial and postglacial periods the Earth experienced ten thousand years ago." As soon as the temperature blipped upward, a tremendous surge occurred in the number of species, a burst of increase in biodiversity.

"Does that remind you of anything?" Jim asked. He knew my interests and was confident of my answer. Well, I ventured, it *looks* like a pattern of stasis and then rapid change, of punctuated equilibrium. "Doesn't it," he said. "And yet conventional evolutionary theory would predict gradual change." If the system had come to rest at the edge of chaos, an environmental jolt would push it into the chaotic regime, I mused, and an avalanche of change, a burst of speciation, would be predicted. Interesting, I said. *Very* interesting. "Now it's time for that walk."

I knew I needed to talk to Stuart Pimm. An ecologist at the University of Tennessee, Knoxville, Stuart recently wrote a book called *The Balance of Nature?*, which effectively sets the agenda for a new understanding of the world of nature. "It's a fuzzy phrase," he said of the title, "but most people understand it to mean something about nature's ability to restore itself after some kind of disturbance. And that ability is thought of as arising from within 'nature,' the ecological processes within populations, among the interactions between species in a community, and between the community and the physical environment."

The phrase has also had mystical connotations, hasn't it? I asked. The idea of nature as some kind of superorganism, almost sentient, and able to heal itself? "That's true. There was a fashion a while back for what I call mystical ecology, the notion of all kinds of emergent properties of nature that you couldn't understand, weren't meant to understand, and if you could, they were considered to be no longer important. I'm suspicious of emergent properties I can't understand."

We were walking in the lower elevations of the Great Smoky Mountains, south of Knoxville. In the early 1800s settlers came to this area and hewed cornfields among stands of ancient deciduous forests. Abundantly watered from the rain-catching peaks, which rise to six thousand feet, the land was productive. Until half a century ago a small community, Forks of the River, thrived here, with twenty-five farmsteads, a church, school, store, post office, grist mill, and sawmill. Since then the land has been part of the Great Smoky Mountains National Park, and evidence of its recent past is difficult for the untrained eye to see, as the forest repairs itself.

I had explained to Stuart my interest in exploring the extent to which the new science of Complexity might be relevant to nature, to the important patterns of biology. The structure and behavior of ecological communities was very much part of this venture, not simply as potential components of Gaia, but in their own right. I asked whether it was reasonable to think of ecological communities as complex dynamical systems. We had stopped by a small creek, which the settlers had used as a means of transport before they put in simple roadways. The nearby trees were small in diameter, a sign that we were looking at land that had once been cleared for farming. Stands of yellow poplar and white pines, and honeysuckle, too, are signatures of the forest in the process of repair. "There's a very long answer to that question," Stuart said, "and we'll go into some of it. But the short answer is an emphatic yes."

That isn't the conventional wisdom of modern ecology, is it? Isn't much of ecology based on the idea of simple equilibria, and that the behavior of species in ecosystems is predictable in that kind of framework? "That's right. But it's clear to me that we have to think of species as being embedded in complex dynamical systems, and this gives you a very different view of the world. For the next five years people are going to tell me I'm dead wrong. Then, when the idea has finally sunk in, they'll say they knew it all along."

Stuart is unusual among academic ecologists. He is a leading theoretician and an enthusiastic field worker. He is also passionate about ecological conservation and restoration. For Stuart, neither theory nor practice dominates his worldview; they meld in a creative union.

I asked him what signatures from ecosystems indicate that the dynamics of complex systems underlie much of nature. "Foodwebs, for instance," he said. "You can view foodwebs as an emergent property of complex systems." Ecological communities may be made up of just a handful of species, or many hundreds, and they may include the range of biological roles: primary producers, such as plants and algae, herbivores, carnivores, omnivores, parasites, and so on, all living in a network of complex interdependence. Darwin, at the end of *Origin of Species*, portrayed the interconnectedness of ecological communities in a famous passage: "It is interesting to contemplate an entangled bank, clothed with many plants of many kinds, with birds singing on the bushes, with various insects flitting about, and with worms crawling through the damp earth." Foodwebs, as they exist in nature, are the result of who eats whom, and as constructed on paper by ecologists represent a roadmap through the entangled bank.

"The remarkable thing about foodwebs is that they have just a few major characteristics," Stuart said, "such as the length of the food chains (a progression of who eats whom, from the bottom of the foodweb to the top), and the ratio of predator species to prey species. You see common patterns wherever you look." Overall there is a balance between the number of species in the community, and the pattern and strength of links among them. The fact that such strong patterns exist where there is the potential for bewildering variety indicates something deep in the organization of ecological communities. As Stuart described it, I was reminded of the order that emerges from Stu Kauffman's Boolean networks. Is there something fundamentally similar about such networks and the order you see in foodwebs? I asked Stuart. "Yes, I think that's reasonable."

As we continued our walk we passed John Ownby's log cabin, which the park service rehabilitated in 1963, using yellow poplar and white pine. A tiny structure, Mr. Ownby had built a lean-to room on one side to accommodate his expanding family in the early decades of the nineteenth century. Nearby is a spring the Ownbys depended on, and around the cabin are tall walnut trees, a source of highly valued nuts. For a brief moment in the forest's history, Mr.

Ownby and his fellow settlers had imposed human ecology on the grander ecological community of an ancient deciduous forest.

Stuart told me more about foodweb structures, how they influence the behavior of individual species, and why trying to predict such behavior is often extremely difficult. "I'll give you a simple example," he said. "Ecologists often study predator-prey pairs, and you would imagine that if you were to remove the predator from the community, the prey would benefit, wouldn't you?" I agreed. "Well, imagine that the predator of prey A also eats a second species, prey B. Now imagine that A and B are competitors; they eat the same leaves, for instance, or nest in the same trees; something of that sort. So, if you now remove the predator, species A may be worse off because it may suffer stiffer competition from species B." That's a simple example? "Yes, mostly the ramifications are much more complex," said Stuart. "As we become more familiar with the real complexities of behavior, we are sensing longer and longer swells that are coming from deep within the foodweb." That's a wonderful image, I said, like boats tossed on a sea driven by unseen but powerful currents. Tell me what patterns you see. "I will, but first we'll drive up to Newfound Gap."

It was early May, and as we climbed to more than five thousand feet we left behind a verdant spring, with the dogwood season losing grip on its last glorious moments, to briefly return to winter. The deciduous trees had yet to leaf, and the only green to be seen was on the conifers, which, sadly, showed signs of damage from acid rain. The view, however, was spectacular, as Stuart had promised, the peaks of the Great Smoky Mountains around us, climax of the Appalachian Highlands. An ardent ornithologist from a young age, Stuart was constantly identifying birds for me, mostly from their song. "Even in this kind of cover it's often difficult to see them."

Stuart told me that during the past decade or so, he and several colleagues had turned from studying the properties of foodwebs to looking at the way they might be assembled in nature. From this enterprise several remarkable insights into the dynamics of ecological communities emerged.

It started when Stuart and Mac Post tried to build ecological communities in a computer model. They added one species at a time (plants, herbivores, and carnivores), each mathematically defined with a small suite of behaviors, such as how much territory each individual typically requires, how much food it needs and of what type, and which other species might be its prey or its predator. Some species succeeded in penetrating the growing ecosystem, while others failed. "We got two results from this," Stuart explained. "First, up to about twelve, more or less any species would succeed, provided it was ecologically sensible—you can't put herbivores in before you have plants, for instance." You mean, a species can successfully invade a community when there are only a few members already in it? "Yes, species-poor communities are easy to invade. The next result has two parts," Stuart continued. Let me guess, I said. Species-rich communities are difficult to invade? "That's right, but it's more interesting than that, and it puzzled us for a long time. We found that newly established species-rich communities are more difficult to invade than species-poor ones, but mature communities are even tougher."

I had read a paper the previous year, by Ted Case, an ecologist at the University of California, San Diego, in which he showed a similar phenomenon in a computer model of an ecosystem. He had written that the interactions among the species in the community create "an invisible protective network" that tended to repel potential invaders. I asked Stuart if this was the same kind of phenomenon as in his earlier models. "Yes it is," he replied. "And the question is, What is the nature of the protective network?"

Before we get into that, I said, these are computer models, right? "Yes they are." Well, I asked, do they match the real world? "I'll tell you about Hawai'i," Stuart answered. For about three months of every year Stuart conducts fieldwork in Hawai'i, high up in the rain forest, where the annual rainfall is a staggering thirty feet. (This put my protestations in Bill Ray's rain forest in Costa Rica in humbling perspective.) Stuart knows the terrain well, especially the three-day hike from civilization to his study site.

"More species of birds and plants have been introduced into

Hawai'i than anywhere in the world," explained Stuart. "But there are two separate ecological worlds. There's the highland region, which is still pristine, with native plants and birds. Relatively few species have invaded here, and this to me represents the persistent community, the long-established community that resists invasion. And there's the lowland region, in which human settlement has disrupted established communities and made them vulnerable to invasion. This is like the immature communities in our models. It's often disconcerting to be walking around in the lush, tropical forest in the lowlands, and you hear this 'swit . . . swit . . . swit,' and you think, 'That must be an exotic bird.' Finally you see perched in a typical tropical forest tree, festooned with lianas and epiphytes, a common cardinal, like the ones we saw earlier today."

Stuart and Mac's community-assembly models consumed huge quantities of computer time, and produced huge quantities of output. "One summer's work would be stacked eight or nine feet high," said Stuart. "We were puzzled why mature communities were more persistent than newly established ones, that they more effectively resisted invasion. We thought maybe that a selection process was going on, with only the better, more efficient species being able to penetrate as time went on." What do you mean by better, more efficient? "Plants that had a higher productive rate, herbivores that could gather more food, carnivores that could run faster, that kind of thing," Stuart answered. "It seemed plausible, but we couldn't confirm it in our data."

Meanwhile, Jim Drake, an ecologist at Purdue University but now a colleague of Stuart's at Knoxville, was working on the same problem. Jim started with a pool of plants, herbivores, and carnivores, 125 species in all, and had his computer pluck single species at a time for possible entry into an assembling community. If a species failed the first time, it might have a second chance. Like Stuart and Mac's model, an extremely persistent community eventually emerged, containing about fifteen species. Jim then went on and discovered two fascinating and important results.

First, if he started all over again with the same original pool of species, he again finished up with an extremely persistent community, but one of a different composition from the first. He ran it a third time, with the same result: a persistent community, different from the first two. "Jim produced many different persistent communities," Stu told me. "And he could tell that there was nothing particularly special about the species in them; they weren't particularly better in any way than their competitors. What was special was the dynamics of the persistent communities themselves. Most species could become a member of a persistent community, given the right circumstances. Now that's about as emergent a property of an ecological community as you can find." The global property of persistence, arising from interaction among species in the community, and not particularly special species at that. Yes, that's a wonderful example of emergence, I said.

"The second result is even more amazing," said Stuart. "Take one of these persistent communities with its, say, fifteen species. Now reassemble the community from the beginning using only these same fifteen species, and you find you can't do it, no matter what order or combination of orders of introduction you try. You simply cannot put the community back together again once you've taken it apart. I call it the Humpty Dumpty effect." Wonderful image, I said. But how do you explain it? "Jim didn't know, and I didn't know. And then Stu Kauffman came to Knoxville to give a talk about his Boolean networks and rugged fitness landscapes, and I said to myself, 'That's it. That's where we'll find the answer.' "

While for Stu, the Boolean networks produced different states of the genome, for Stuart they became the presence or absence of species in a community. "It was an intellectual leap for us, not big, but crucial," said Stuart. "Very quickly we could see how mind-bogglingly complex our communities were, in terms of transitions from one state to another." The breakthrough came when Stuart saw from the Boolean network analysis that some of the transitions you'd want to make in order to reassemble a persistent community simply could not occur. It was the old "you can't get there from here" problem, written in higher mathematics and illuminating a major ecological conundrum. "I ran down the corridor, banged on

Jim's door, and yelled, 'I've found Humpty Dumpty,' " recalled Stuart, with a mixture of mirth and triumph.

It was a thrilling idea, but difficult to grasp. I asked Stuart if he was saying that persistent communities can be assembled only if, on the way to them, other species come in and out of the community, like stepping stones to a more stable state? "Yes," he said. "But let me give you another image. See the peaks around us?" Yes. "See how between the peaks you can't see the valleys, because of mist and low cloud?" Yes. "Well that's how our assembly looks." Nice image, I conceded, but what does it mean?

"When we stripped down our community assembly model and used Stu Kauffman's rugged landscape idea, we found several surprising things. First, even random assembly will give us a degree of order I simply would not have expected. And second, the communities behave as if they were climbing adaptive peaks. We can't say much about what's happening in the valleys, but higher up, that's how it looks. That's the image of the Smokies here, peaks protruding through the clouds." Is it legitimate to talk about communities climbing adaptive peaks, becoming fitter? "No, because we can't say what fitness for a community is," Stu replied. "But instead of wandering around lost, the communities quickly climb peaks, and that represents the persistent states, many of them." And once you're on a local peak, you can't easily get to another one? "That's right. Humpty Dumpty lives!"

I had hoped to find some imprint of the dynamics of complex systems in ecological communities, but I was not prepared for what Stuart had told me. Emergence was everywhere, not mystical but as a result of local interaction. As Stuart said, the answer to my original question was an emphatic yes. Community ecology has proved to be notoriously difficult to unravel by conventional analysis, and the reason is now obvious: complex dynamics are tough to penetrate. But once you begin to view ecological communities in the context of complex dynamical systems, patterns appear.

There was one more pattern I wanted to ask about. You've talked to Stu Kauffman, I said. You know his interest in the edge of chaos and self-organized criticality. Do you see any sign of that in your systems? "Not long ago Jim Drake showed me some data on what happens when he perturbs his persistent communities," replied Stuart. "I looked at the data, which showed a whole range of extinction events, and said, 'Jim, have you heard about self-organized criticality?' He hadn't. I said, 'I bet if we plot these things we'll get a power law.' " Stuart was right. That means the connectedness within communities has to be considerable, doesn't it? "It has to be, otherwise extinction avalanches wouldn't propagate through them." Stu Kauffman would be pleased.

I asked Stuart what this said about real ecological communities. "It's hard to get the data you'd like from natural communities," lamented Stuart. "But we can say that the foodweb patterns we see in our community models look very much like the ones that have been drawn for natural communities, so you could take that as suggestive." Suggestive that natural communities during assembly bring themselves toward a critical state, the edge of chaos? "If you forced me to be definitive, I'd have to say yes."

With that we began our departure, walking again through stands of pine blighted by acid rain. "Who knows where it comes from," sighed Stuart. That example of inimical interaction between physical and biological worlds before us reminded me of a sentence I'd recently read in one of Stuart's papers. When you said, "There isn't a goddess Gaia," what did you mean? "I meant that there's no external 'something' controlling global ecology."

The paper reports the new work on community assembly on rugged landscapes, in which Stuart and his colleague Hang-Kwang Luh write: "It looks *as though* there is something that pushes the assembly towards peaks in the landscape and *as though* there is something *akin to* fitness." So you're saying that whatever the communities do, they do as a result of internal dynamics, not in response to anything external? "That's what we're saying." An emergent property of a dynamical system? "Yes." Can you conceive of emergent properties on a global scale that could produce homeostasis of the Gaia variety? "If that's what Lovelock has been saying, it's been pretty obscure to me," replied Stuart. "I heard him recently and he really did seem to come close to the edge of an appeal to mysticism."

I told Stuart that I had recently visited Lovelock and was convinced that whatever mysticism was associated with him was a result of the translation of the message, not the message itself. I also said that the kind of emergent properties that he, Stuart, and his colleagues were uncovering in ecological communities seemed to be of the character of mechanism that Lovelock had in mind, when linked with physical systems. Or as Stu Kauffman might put it, individual entities in the system myopically pursue their own ends, with collective benefit being the result.

Why don't you contact Lovelock? I suggested to Stuart. You might be surprised. "OK," he replied. "I might just do that."

Coombe Mill is situated on thirty-five acres of land, long and thin, with a mile of the River Carey running through it. Since moving there fifteen years ago, Jim and his family have planted more than twenty thousand trees, ash, elder, beech, and oak. The intention is to restore the land to what it was like before Iron Age deforestation denuded the entire region, including Dartmoor. "We can walk three and a half miles if we go up there, by the old railway track, and back by the river," said Jim, displaying the pleasure he takes in his daily ramble.

I asked whether he was concerned that people reacted negatively to the way Gaia is often discussed. "There's a huge amount of literature that is supposed to be about Gaia, that New Age stuff. It's 100 percent rubbish. But you don't mean that?" No, I said. I mean your books and articles, the material from which people have inferred a purposefulness in Gaia. "I recognize I use words that sometimes irritate biologists," he began. "Biologists have fought long battles against vitalism, animism, anything that smacks of some kind of force beyond the immediate mechanics of the system. So, anything that sounds holistic—a dirty word in itself—is viewed with suspicion. I don't have an instinctive reaction against words like that."

We had reached the old railway track, long since devoid of the rails themselves. In England, as everywhere, railway tracks are host to wildflowers that have disappeared from other locations. Unfortunately, February in England is too early for wildflowers, except for some gorse, brilliant yellow splashed against dark green foliage. "You know the country saying about gorse?" Jim asked. "When gorse is in flower, 'tis the season for kissing." He laughed. "Gorse is *always* in flower.

"But you asked about language and Gaia," Jim continued. "Let me tell you a story. A few years ago there was a debate at the Linnaean Society. I was speaking for Gaia and Brian Clark was against. Brian is biological secretary of the Royal Society. We said our pieces, the vote was taken, and it came out in favor of Gaia, even though the audience was mostly biologists. Brian said to me afterward, 'I'd like to know what you're talking about, but you don't speak our language.' " Jim paused, smiled, and said, "One day he will."

C H A P T E R   S E V E N

Complexity and the Reality of Progress

The Villa Serbelloni is situated on a promontory that reaches into Lake Como, Italy. With a history that goes back to the first century A.D., when Pliny the Younger had a villa here, the location has a magnificent view of the lake, enhanced by the rising peaks of the nearby Alps. In 1959 the present villa was bequeathed to the Rockefeller Foundation, which uses it as a conference center.

Conrad Waddington—sensible man that he was—chose the venue for his annual workshops on the emerging theoretical biology, in the late 1960s and early 1970s. It was during one of those conferences, in 1968, that Stu Kauffman gave the first public presentation of his random Boolean networks, his discovery of "order for free." It was here, too, that Stu was offered his first faculty position, at the University of Chicago, on the strength of that talk. And the manicured slopes of the extensive lakeside grounds were the site of yet another Kauffman triumph: he won the paper airplane contest.

"A bit of a cheat, really," Stu admitted to me, chuckling at the memory. "There we were, intensely discussing all these wonderful new ideas, and in the middle of it I threw out the challenge: whose paper airplane can fly furthest?" To John Maynard Smith, who was an aeronautical engineer before he became one of the world's leading evolutionary theorists, the challenge was irresistible. To Morel Cohen, theoretical physicist of exquisite insight, the prize seemed already won: he would work out a design from first principles. To Lewis Wolpert, developmental biologist of world renown, no challenge of any sort could be left unanswered. Richard Lewontin joined in, as did Richard Levins, and several more. "What they didn't know," said Stu, "was that I'd spent years perfecting a design. Started when I was eight."

Which design did you use? I asked, warming to the topic at hand. The straight wing or the delta wing? "The delta wing," Stu replied. We were talking in his office at the University of Pennsylvania, and he scrabbled among the piles on his desk, retrieved a piece of paper deemed suitable to the task, and began folding it in the pattern known to generations of small children. The first fold is lengthways, in half. The second makes a 45-degree angle; the third a sleek, 22½-degree angle. Fold the wings down, and the basic delta wing dart emerges. "My key modification was to bend the point underneath, about an inch, so you finish up with a blunt nose," Stu confided, as pleased with this invention as he might have been for a discovery in the lab next door. "Gives it weight," he explained. "I used to spend hours up at Lake Tahoe, shooting these things into the sky." With that he launched the plane across the office, and sure enough it traveled smoothly, at a low angle, clearly destined for a long flight—until it crashed into an ancient filing cabinet. "We should have taken it outside."

Our discussion had been sidetracked onto the aerodynamics of paper airplanes for good reason. We had been talking about complex dynamical systems, including biological systems, and how they often generate order. We had got on to the history of such systems, of how they change through time. I had noticed on a number of occasions that, when talking about model evolutionary systems, such as Tom Ray's or Kristen Lindgren's, people would frequently refer to the tendency of such systems to generate increasing complexity. You start with a simple system, allow the fundamental dynamics to operate, and products of increasing complexity emerged. It was the nature of mathematical models of complex adaptive systems. It happened in the real world of biological systems. That, unmistakably, repeatedly, was the message.

You're a biologist, I said to Stu. You're aware that your colleagues have a problem with the notion of complexity, not being certain how to define it, not being certain what it really means. "Yes, I know," he replied, now thoughtfully examining his plane and making small adjustments. "I also know that biological systems can't avoid complexity; it emerges spontaneously. And complexity does *seem* to increase through time." He told me that if two Boolean networks interact and play games with each other, they become more complex, and get better at each interaction. He reiterated what others said about Kristen Lindgren's Prisoner's Dilemma model, that the strategies get more complex, get better at the game. "And we all have the sense that biological systems become more complex through time," he added. Get better? I asked. "Well, that's where it gets to be tricky. Look at this plane. Bending the nose under like this—making the design more complex, if you like—makes it fly further. That sounds like 'better' doesn't it? But it can't do aerobatics. Some of those planes at Villa Serbelloni did wonderful aerobatics. So, it depends what you mean by better."

I realized that we were entering difficult territory, one that can sometimes sound like semantic contrariness. More ordered, more complex, better—are they the same thing? Is it even an accurate description of what happens in biological systems through evolutionary time? "It's a profound question," said Stu. "There's a price to pay in becoming more complex; the system is more likely to break, for instance. We need a reason why biological systems become more complex through time. It must be very simple and it must be very deep." You *are* assuming an inexorable increase in complexity? Stu thought again for a few moments, the plane poised for flight. "I am," he said, but with a distinct note of caution. Launched, the plane repeated its steady flight path, and again crashed into the filing cabinet.

"Talk to Dan McShea," I was urged repeatedly. I had decided that I needed to look more closely at biologists' views of complexity. It was clear to me that if the ideas of the Santa Fe Institute were to mean anything in the biological world, there had to be some conceptual common ground between the two. Where would I find it? I contacted several friends, biologists who over the years I had found to be reflective about the larger problems of the science. What is biological complexity? I asked. "If anyone can help you answer that, it's Dan McShea."

Dan, who is now at the University of Michigan, Ann Arbor, was a student of Dave Raup's at Chicago. He had done a study on the incompleteness of the fossil record, an issue that deeply taxed Darwin and remains a matter of profound practical importance to modern paleontologists. Then, in the summer of 1985, a friend gave Dan a book, *The Recursive Universe*, by William Poundstone. "It completely changed my intellectual focus," Dan told me. The book is ingeniously structured around ideas of cosmology and Conway's Game of Life. The cosmology was fascinating, but the Game of Life triggered something deep in Dan. A reflective person by nature, he was already interested in pattern and complexity in nature. The fact that so much complexity flowed from so simple a set of rules, as happens in the Game of Life, was a profound insight for Dan, as it has been for many people. "Because Chicago is heavily theoretical, this experience became transformed into a new research project for me: What has complexity done in the history of life?"

The first thing to be done was a search through the literature, both modern and historical. "I quickly learned two things," said Dan. "First, there is a general though vague consensus that complexity has increased through evolutionary history. Second, *complexity* is a very slippery word. It can mean many things." One of the things with which the word is often conflated, for instance, is "progress," the notion that evolution proceeds along a path toward inevitable improvement. These days biologists are very uncomfortable with the idea of progress, because of connotations of an external guiding force. "It's acceptable to talk about complexity," explained Dan, "but not progress."

The image of an ordered world, with organisms arranged from the "lowest" to the "highest" forms, goes very deep in our culture. It is to be found in Plato and implicitly in the order of creation in Genesis. Much later, in the seventeenth century, this ordering of nature became encapsulated in what was known as the Great Chain of Being. In those pre-Darwinian times, the chain was meant as a static description of each species' place in the world, not as a record of change over time. Humans, not surprisingly, were placed near the head of the chain, "a little lower than the angels." With the advent of Darwinian evolutionary theory, organisms came to be seen as the product of change over very long periods of time. The static order of the Great Chain of Being effectively became transformed in people's minds into a record of that evolutionary history, from simple to complex forms. An increase in complexity through evolutionary time seemed evident.

"Darwin believed it, as did most of his contemporaries," said Dan. "And so did most of the Anglo-American paleontological community, from the last decade of the nineteenth century through to the middle of this one. Then some doubts began to creep in, but it's fair to say that most people still believe it to some degree." Maybe that's because it's true, I ventured. "I have the feeling that many people would like to think it's true, but there is very little solid evidence," replied Dan. "But first, you have to be very clear what you mean when you talk about complexity. It's very easy to begin a conversation with an agreement that complexity is X, and moments later to hear yourself arguing that such-and-such is not complex because it's not Y." Can you be more specific? "In my research I've focussed on morphological complexity, the details of anatomical structure. But I suspect, for instance, you're interested in something more general, something that includes behavior."

Dan was correct. My notion of complexity was inchoate, but I was aware that behavior was a part of it. I've watched troops of vervet monkeys in Kenya, and I don't have any difficulty in thinking of them as a more complex form of life than the trees that are so important a part of their daily lives. The vervets, singly but particularly as a socially interactive network, look more biologically complex than the trees. They also appear to me more behaviorally complex than the zebra and wildebeest that herd nearby.

OK, I said, suppose we put behavior to one side for a while, how do you approach morphological complexity? "You also have to put any notion of 'better' to one side," cautioned Dan. "That really is an elusive idea. You could say that an increase in the number of components represents more complex—and better—forgetting that a sun dial is likely to break down less frequently than a watch." A few biologists have tried to pin down criteria for measuring complexity, including the number of different anatomical parts, with only modest success. Is a cat more complex than a clam? It would be judged so by this criterion, but is it true in an absolute sense? And am I being unfair to trees in thinking vervets more complex? After all, I can empathize with the life of a vervet, but not a tree. Perhaps I'm missing something about the complexity of tree-ness?

There is a natural tendency also to think of mammals as somehow more complex than reptiles. A lion seems a more advanced machine than, say, a tyrannosaurus, even though both are (or were) carnivores. Zebras are surely more complex in some way than hadrosaurs, even though both are (or were) grazers and social animals. But one thing that has become clear to biologists recently is that the modern world of mammals is just like the ancient world of the great reptiles. In both you can identify small and large carnivores, small and large herbivores, small and large insectivores, and so on. The same ecological niches are filled in both worlds, and with about the same number of species. Nothing to distinguish them there. But are mammals, with their higher metabolic rate, more complex than reptiles in a general sense, because they channel more energy? Some dinosaurs were probably warm-blooded, too, so that notion is not clear-cut either. "If we are going to get anywhere with this, we need to focus on something discrete, something measurable," said Dan.

One of the most respected attempts for an objective measure of complexity was developed by John Tyler Bonner, of Princeton University. Count the number of different cell types in the organism, he suggested. In principle this gives a sense of the number of specialized functions an organism can perform, and that smacks of complexity. It also has the virtue of considering the whole organism, not just one part. (It does leave out behavior, but we are only considering morphological complexity here.) Bonner was able to show higher complexity in larger species by this measure, but he did not try to determine whether it increased through evolutionary time. That, however, would be a reasonable inference. All this recalled something Edward O. Wilson once said to me: "It is not difficult to recognize complexity, Roger. The difficulty comes in how you measure it."

"Let me show you what I tried," said Dan. From a large drawer he pulled out a small skeleton, a squirrel. "I decided I would look at complexity in vertebral columns," he explained. "If you look at a fish's vertebral column, all the vertebrae are virtually identical. Anything else, a mammal for instance, is more complex than that: you get different structures in the cervical, thoracic, and other regions. So that looks like an increase in complexity, from fish to, say, squirrel, doesn't it?" I agreed it did. "Now let's look at modern squirrels and their ancestors. If there's a trend to more complexity, you'd expect to see more complex anatomy in modern species, wouldn't you?" Again I agreed. "I measured various aspects of the vertebrae along the column, in ancestor and descendants. I did this for squirrels, ruminants, camels, and a few others, and then devised three metrics, three ways to capture the morphological complexity in the vertebrae of ancestors and descendants." He turned to a pile of papers, pulled out a sheet. "Look at this."

In front of me was a table of data, showing the three types of animal and the various complexity measures in the vertebrae. The table was full of Ds and Is, for decrease and increase in complexity. "It's amazingly equal, the same number of decreases as increases," explained Dan. It's not going anywhere, I said, no evidence of increasing complexity? "None. True, it's only a short period of time, about 30 million years. But you might have expected to see some tendency toward more complexity, if that's what happens in evolution."

We had been switching from "What is complexity?" to "Does it increase through evolutionary time?" with alarming ease. Dan was right when he said *complexity* was a slippery word. The whole issue seemed to be slippery. And, I have to admit to feeling "So what?" about Dan's result on the vertebral column. I wondered what we could have said if the data had shown an increase in the complexity measure he chose. "That over this period of time vertebral columns *had* increased in complexity," Dan replied. But it might be the only thing that had increased, and that wouldn't be very profound, would it? "You're right. And I think you're getting to see how very difficult this whole business is."

Dan was right. I had the sense of complexity as a mirage: I was certain of its existence, until I tried to reach out for it, tried to anchor it in reality. Where, I wondered, would biological complexity be anchored by someone with a Santa Fe Institute view of the world? "I don't see the problem," Norman Packard told me. "Biological complexity has to do with the ability to process information. Computational capability, that's what we see in our cellular automata models, and in other complex adaptive systems. I view organisms as complex dynamical systems, and what drives their evolution is increased computational ability."

But, I asked, is it really valid to describe what you see in evolutionary models as an increase in complexity? "I don't know what else you'd call it," said Norman. "Look, with Kristen Lindgren's model you start with the simplest possible strategy and you finish up with complex individual strategies and a complex interactive system. And it's simply the dynamics that produces it, given the goal of playing the game. You see the same kind of thing in Tom Ray's model, and that's even more like biological evolution." But some of Tom's critters become simpler, I reminded Norman, some shorten their code and become parasites, about half the size of the ancestral organism. "That's true, but, first, I'm not saying that every organism need itself become more complex; second, the system as a whole undoubtedly becomes more complex, no question about that. You've seen it. You know what I mean."

Norman's argument struck a chord with several comments I'd come across among biologists. For example, in a classic 1977 text on evolution by Theodosius Dobzhansky, Francisco Ayala, G. Ledyard Stebbins, and James Valentine, the "ability to gather and process information" is said to have increased through evolutionary history, and, indeed, to be a mark of progress. A few years ago I attended a conference at the Field Museum in Chicago, where the topic was "evolutionary progress." Francisco Ayala was one of the first to speak. "The ability to obtain and process information about the environment, and to react accordingly, is an important adaptation because it allows that organism to seek out suitable environments and resources and to avoid unsuitable ones," he said. Ed Wilson also considers information processing as a measure of complexity. "No question about it," he told me. "There's been a general increase in information processing over the last 550 million years, and particularly in the last 150 million years." If at least some biologists and the dynamical systems people collectively point to information processing as a mark of complexity, we may be getting somewhere.

I can see what might be meant by computation in organisms that have a brain of reasonable size, I said to Norman, but what about clams and trees? "Survival has to do with gathering information about the environment, and responding appropriately," Norman answered, clearly echoing Ayala. "Bacteria do that, by responding to the presence or absence of certain chemicals and by moving. Trees communicate chemically, too. Computation is a fundamental property of complex adaptive systems, which, you'll remember, is optimized at the edge of chaos. Any complex adaptive system can compute; that's the key point. You don't have to have a brain to process information in the way I'm talking about it." But it helps? "It's higher on the scale of computational ability, if you like."

The phrase "higher on the scale" is instantly provocative to biologists, because, with Darwin cited as the example, they are taught that "higher" and "lower" are value-laden terms, not meaningful biological terms. They are also taught that higher and lower imply a progressive element in evolution, ascending the scale of nature from the simple to the complex. As Dan McShea said, biologists are willing to tackle the notion of complexity and accept that it has increased in the history of life in some ill-defined way, but to speak of "progress" is regarded as unwise. If evolution is said to be progressive, then it is all too easy to see it as being directed, following an arrow of improvement. And that is all too redolent of the Divine design of pre-Darwinian days.

When you say "higher on the scale," I asked Norman, are you suggesting a history of successive increases in computational ability in evolution? "That's how it looks to me," he replied. "Intuitively, it seems reasonable that the task of survival requires computation. If that's true, then selection among organisms will lead to an increase in computational abilities. That creates an arrow of change, not just a drift upward." Can't you be accused of being anthropocentric, I said, viewing the world from this pinnacle of computational power we have in our heads? "Humans stick out like a sore thumb, with our relatively enormous brain, but if you leave us out of the equation it is still correct to say that computational ability has increased through time, and that's just what you'd expect."

Most species on Earth today are single-celled organisms, as in the pre-Cambrian, and much of the rest are insects, I said. That doesn't look like inexorable progress to greater computational ability, does it? "We're talking about survival," said Norman. "And, yes, there are countless niches out there in which species do very well with certain levels of computational abilities. But where survival is contested, mostly you will see an increase. Think of it as a constant exploration of the utility of increased computational complexity in evolution. Sometimes it gives an advantage, and that's what gives you the arrow."

I asked Norman if he was aware that most biologists would be uncomfortable with the kind of progress he sees in evolution. "People don't like it for sociological, not scientific, reasons," he responded. "I don't impute a value judgement to computational superiority."

"Progress is a noxious, culturally embedded, untestable, nonoperational idea that must be replaced if we wish to understand the patterns of history." With that statement Stephen Jay Gould opened his presentation at the 1987 conference on evolutionary progress, at Chicago's Field Museum. More strongly worded than most, Steve's argument nevertheless characterized the sentiment of the day. With the exception of Francisco Ayala, who tentatively admitted that given certain caveats he could see progress in evolution, speaker after speaker denied its existence. Why, I asked Steve, do you consider progress to be noxious? That's a strong word.

I realized I needed a clearer sense of biologists' abhorrence of the notion of progress if I was going to create a clear vision of the Santa Fe Institute's view of biological evolution. I have visited Steve many times at the Museum of Comparative Zoology, Harvard University. Located in the oldest wing of the museum, his "office" is a corner of an enormous room that is divided by tall collection cabinets, many of which contain thousands of the shells of *Cerion*, a West Indian land snail, Steve's favorite organism of study. Bookshelves mark something of a boundary to the office area, and the quick eye can spot Victorian editions of Darwin, Thomas Henry Huxley, Herbert Spencer, and Comte Georges de Buffon (thirty-one volumes), among many others. The faded green paint on the walls is also Victorian, and bears the calligraphic script of an exhibition of that era's view of the biological world. "Synopsis of the Animal Kingdom," reads one notice, just disappearing behind high storage cabinets. Other labels—such as "Sponges and Protozoa," "Mammals," and "Vermes" (meaning worms)—are to be glimpsed here and there, often partly hidden behind shelves and desks. Steve usually sits in an old cane chair, the stuffing of which is breaking out of the cushion. Today, however, he's at a desk, soon to depart to give a lecture.

"Progress is not intrinsically and logically noxious," he replied. "It's noxious in the context of Western cultural traditions." With roots going back to the seventeenth century, progress as a central social ethic reached its height in the nineteenth century, with the industrial revolution and Victorian expansionism, Steve explained. Fears of self-destruction in recent decades, either militarily inflicted or through pollution, have dulled the eternal optimism of the Victorian and Edwardian eras. Nevertheless, the assumed inexorable march of scientific discovery and economic growth continue to fuel the idea that progress is a good and natural part of history. "Progress has been a prevailing doctrine in the interpretation of historical sequence," Steve continued, "and since evolution is the grandest history of all, the notion of progress immediately got transferred to it. You are aware of some of the consequences of that."

One consequence was that evolution, in being viewed as progressing from lower to ever higher forms, was considered to lead unswervingly to the emergence of humans. This has been expressed openly in early writings, by Robert Broom, for instance, a paleontologist who discovered many ancient human fossils in South Africa in the 1940s and 1950s. "Surely there can be no subject so interesting to Man as why he has appeared on earth," he wrote in 1933. "Much of evolution looks as if it had been planned to result in Man, and in other animals and plants to make the world a suitable place for him to dwell." While few went as far as Broom, many promoted the notion of the inevitability of *Homo sapiens*. "Life, if fully understood, is not a freak in the universe—nor Man a freak in life," Pierre Teilhard de Chardin, philosopher, anthropologist, and Jesuit priest, wrote four decades ago. "On the contrary, life physically culminates in Man, just as energy physically culminates in life."

The same notion is alive and well today, living in the *New York Times*, albeit expressed in less florid language than Teilhard's. Reporting the discovery of an even earlier ancestor of vertebrates than had been known previously, a news report described vertebrates as "the group that *led to* humans." (My italics.) Perhaps it was a news editor's infelicity; perhaps the reporter really meant to say that *Homo sapiens* is the culminating product of vertebrate evolution these past 550 million years. One of the dozens of species of cichlid fish that have evolved in Lake Victoria within the last few thousand years may seem to have a better claim than *Homo sapiens* to being the end product of vertebrate evolution, having arrived on the scene much more recently.

A second consequence, related to the first and deserving to be termed noxious, is racism, which appeared explicitly in anthropological literature of the turn of the century. "The notion of progress in evolutionary history made easy the acceptance of one race dominating another," said Steve. British and American scholars of the period viewed human evolution as progress through effort on the part of our ancestors (thus nicely reflecting the Victorian work ethic). Our ape cousins were left behind in biological obscurity, victims of their indolence. It also meant that some "races" of humankind fared better than others through their own endeavor: there is no prize for guessing who was first and who last on this scale.

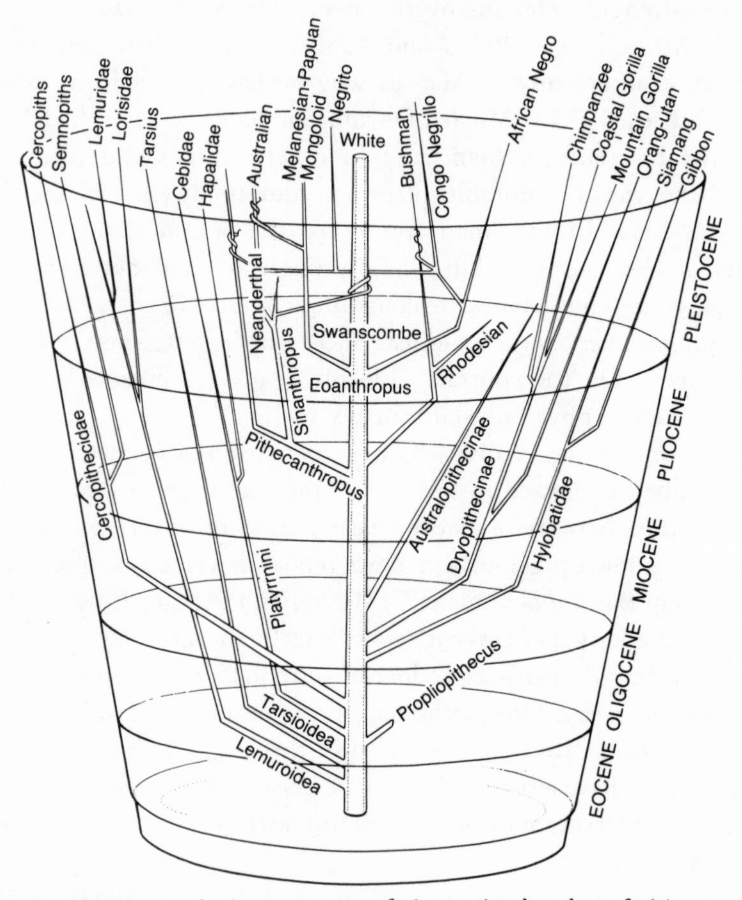

Fig. 7. Human evolutionary trees of the early decades of this century frequently portrayed "whites" as the central, most advanced "race," as seen here in Earnest Hooton's *Up from the Ape*, 1946.

Pictures of the human family tree clearly showed this scale of supposed superiority among races. And the words of dozens of respected anthropologists proclaimed it: "Darwin's doctrine of evolution . . . has been, and ever will be, the means of progressive evolution," wrote Henry Fairfield Osborn, director of the American Museum of Natural History in the early decades of this century. "To extinguish the spirit of competition is to seek racial suicide." Osborn's colleague Roy Chapman Andrews penned similar sentiments: "The progress of different races was unequal. Some developed into masters of the world at incredible speed." Such were the statements of the leaders of the profession, not an extreme fringe. Modern evolutionary texts contain nothing like them. Nevertheless, said Steve, "There is a profound unwillingness to abandon a view of life as predictable progress, because to do so would be to admit that human existence is nothing but a historical accident. That is difficult for many to accept." Progress gives meaning to life.

Just because a scientific idea is imported into social values—however improperly used—doesn't invalidate the original idea, I said to Steve. "Of course not. But global progress is *not* a consequence of the mechanics of natural selection. Darwin recognized that, and that's why he wrote to Hyatt, 'After long reflection I cannot avoid the conviction that no innate tendency to progressive evolution exists.' " Steve has the enviable ability of quoting at length from memory, as he did in this case. Alpheus Hyatt, an American biologist, corresponded with Darwin during the 1870s, and this phrase of Darwin's from one of the letters has become famous. It is also controversial.

"You have to understand that Steve's position on this issue is deeply ideological," Robert Richards told me. "He is not alone among modern evolutionary biologists in denying progress, but he is among the most vocal." Bob, a philosopher and historian of science at the University of Chicago, has recently completed a scholarly book, *The Meaning of Evolution*, in which he argues that Darwin did indeed view evolution as progressive. "Look at this," he said, flicking through his book for a passage from one of Darwin's early notebooks. "Darwin wrote, 'The simplest cannot help become more complicated; and if we look to the first origin, there must be progress.' That's pretty clear, isn't it?" I had to agree it was. But, I countered, I'd also seen anti-progress comments by Darwin. "I'm sure you have; there are quite a lot. You can play the pro-progress versus anti-progress quotation game all day long. But my view is that on balance Darwin is a progressionist." So why is Steve Gould so vehement in opposing progress? I asked. You said it was ideological.

"If you want to see where Steve's rejection of progress comes from, just read that," said Bob, handing me a copy of his book. It was open at a quotation from a book of Steve's, in which he points to late-nineteenth-century Germanic ideas of progress in evolution as contributing eventually to the rise of Nazism. "You can see why Steve is unable to think that Darwin embraced these same ideas of progress," said Bob. "His rejection of the idea of progress, both for himself and for Darwin, is influenced by ideology."

I asked Steve whether his views were ideologically influenced. "Whose aren't?" he responded instantly. "But if you want to know what Darwin's views were, read the *Origin*. It's full of statements to the effect that his theory does not lead to overall progress." That's true, I conceded, but Darwin also seems to admit progress in many places, doesn't he? "That's right," said Steve. "The most famous one comes from near the end: 'And as natural selection works solely by and for the good of each being, all corporeal and mental endowments will tend to progress towards perfection.' But historians make a mistake when they try to find utter consistency in the world of great thinkers. There was a schizophrenia in Darwin, a duality: on one hand he was a philosophical radical in many things, and on the other a comfortable Victorian gentleman, living in a nation in which progress was as intrinsic a presupposition as in any culture in history. But the bare-bones mechanics of his theory of natural selection make no statement about progress. He's very clear about that, he revels in that, and that's why he says 'never say higher or lower.' "

Natural selection concerns simply the adaptation to local circumstances, continued Steve, and as such contains no tendency to global progress. The environment changes in one direction, and adaptation tracks it. The environment changes in another direction, and adaptation tracks it again, blindly and with no direction. With Norman Packard's notion of an inexorable rise in computational ability in mind, I planned to ask Steve about brains. The fossil record shows a dramatic increase in average brain size with the evolution of mammals from reptiles, some 230 million years ago; a similar increase occurs when "modern" mammals evolved, 50 million years ago; and primates are twice as "brainy" as the average mammal (humans, as Norman Packard said, stick out like a sore thumb and are best left out of the equation). Doesn't this mean something? I asked Steve. Doesn't it show an increased ability to process information?

"Look, forty thousand species of vertebrates, right? About twenty-five thousand are fishes . . . no trends there. So, you start with 55 to 60 percent of vertebrates with no trend to bigger brains. Then you have eight thousand species of birds . . . again no trend to bigger brains since their origin. Six thousand species of mammals, a fraction of all vertebrates, and, yes, you do see trends in some groups. So, are you saying that what happens in some groups of six thousand species of mammal represents the thrust of evolution?"

A tough case to answer. Nevertheless, I said, there *are* trends and it's hard to ignore the effect of bigger brains. I find it difficult not to think that it represents something creative in evolution, some measure of increasing complexity. Don't brains embody a higher level of complexity than, say, the structure of the skull or feathers? "Certainly brains have had more effect than any other structure," said Steve. Isn't that a legitimate measure of complexity? "Oh no, because probably next in effect are the bacteria. Effect has to be divorced from complexity." Like Dan McShea, Steve seemed determined to exclude aspects of behavior from complexity.

"The not-so-hidden agenda in all this is a concern with human consciousness," said Steve. "You can't blame us for being fascinated with consciousness; it's an enormous punctuation in the history of life. I view it as a quirky accident, but most people apparently don't want to look at it like that. If you believe there is an inexorable increase in brain size through evolutionary history, then human consciousness becomes predictable, not a quirky accident. Ours is a very 'brain-centric' view of evolution, a bias that distorts our perception of the true pattern of history."

I found many biologists distinctly uncomfortable with talking about increase in brain size as a measure of complexity. "I'm hostile to all sorts of mystical urges toward greater complexity," said Richard Dawkins when I asked him whether an increase in computational complexity might be considered an inevitable part of the evolutionary process. "You'd like to think that being able to solve problems contributes to Darwinian fitness, wouldn't you?" said John Maynard Smith. "But it's hard to relate increased brain size to fitness. After all, bacteria are fit." Michael Ruse, a philosopher of science at the University of Guelph, Canada, who told me, "Scratch any evolutionary biologist and you find a progressionist underneath," also equivocates on this issue: "Can you really say a brain is better than a shell?" Ed Wilson, however, had no doubts: "Braincentric?" he laughed. "Isn't that the ultimate politically correct mode of reasoning? . . . Need I say more?"

By now it had become clear that if Norman Packard is correct in suggesting that an increase in computational ability represents an arrow in the evolutionary process, many biologists will have problems in coping with the message that the new science of Complexity may be bringing them.

When I had talked with Dan McShea, he told me that although there were few hard data on the question of generating biological complexity, there was no shortage of theories. "You can describe most of the theories as either internalist or externalist, and some are more mystical than others." Mystical? I said. "That may be unfair, but you'll see what I mean." For instance, Jean-Baptiste de Lamarck, whose pre-Darwinian theory of evolution greatly influenced Darwin, believed that organisms responded to an innate drive to greater complexity, mediated by invisible fluids. Internalist and mystical? "Definitely mystical," said Dan.

"Spencer, however, is more interesting." Herbert Spencer, nineteenth-century English intellectual, known as much for his leaden prose as his radical social theories, was tremendously influential in his day. He constructed grand syntheses among science and nature, society and psychology, and coopted Darwin's theory of natural selection as a theory of social systems. His was the phrase "survival of the fittest," and his theory became known as Social Darwinism. "Progress . . . is not an accident, but a necessity," he wrote in 1851. "Instead of civilization being artificial, it is a part of nature; all of a piece with the development of the embryo or the unfolding of a flower." Spencer's high social standing and intellectual influence were matched only by the rapidity of his fall into disfavor with the rejection of Social Darwinism. "You barely see his name mentioned today," said Dan. "Either people have forgotten about him or dare not mention him, for fear of being tainted."

Why mention him now? "Spencer had this grand theory—all his theories were grand—about the condensation of order from disorder, heterogeneity from homogeneity as he called it," Dan explained. "He said that dynamic systems have a tendency to become more concentrated and heterogenous as they evolve. He called it the Law of Evolution." By heterogenous, do you mean structure, useful order? "Spencer was talking about all dynamical systems, not just biological systems—physical worlds, biological worlds, and social worlds." The formation of stars, of biological form, and of complex societies? "Yes." He sounds ahead of his time, I commented. It's very reminiscent of the kind of thing the Santa Fe Institute people would say: order crystallizing out of chaos. Is that a fair comparison? "It's exactly what Spencer was saying: consider a homogenous system governed by simple rules or forces. If you just jostle it, heterogenous structure will emerge. Spencer says that the simple, or homogenous, system is unstable—like a balanced scale, it inevitably becomes unbalanced due to rust, wind, and so forth."

Spencer's is an internalist theory of complexity, and, though a little mystical, too, is something of an intellectual antecedent to the science of Complexity. Many great ideas have antecedents, in spirit if not in fact. But Spencer's Law of Evolution is missing something, because the new science of Complexity includes external as well as internal factors. The external factor is selection.

Natural selection would be considered an externalist mechanism for generating complexity, wouldn't it? I asked Dan. "Among several others, yes." Darwin's metaphor for the effect of natural selection is the wedge, crystallized in a famous passage in *Origin of Species*. He imagines the biological world to be packed with species, and the only way for a new species to succeed is by dislodging an incumbent: "Nature may be compared to a surface covered with ten thousand sharp wedges . . . representing different species, all packed closely together and driven by incessant blows . . . sometimes a wedge of one form and sometimes another being struck; the one driven deeply in forcing out others."

Competition abounds, each species jostling with ecological rivals. It is easy to imagine one species gaining a slight advantage, and then its competitors struggling to catch up. This is Stu Kauffman's example of the frog and the fly, writ large. In the end, each species may be improved—that is, be faster, tougher to eat, or smarter than it once was—but none would have achieved an absolute advantage over the others. Progress has occurred (if we may use that word), species may be better at what they once did, but none is any better off. Leigh Van Valen's Red Queen effect—all species running continually to remain in the same place—is a popular image. So is an arms race, for obvious reasons. Whatever we call it, the effect represents a process by which complexity—by some measure—is increased, and we can see that it is driven externally.

If arms races are common in biological systems, then the opportunities for exploring the "utility of increased computational complexity in evolution"—Norman Packard's phrase—would also be common. And the ability to do this would be a significant landmark on the evolutionary landscape. In *The Blind Watchmaker*, Richard Dawkins seems to indicate that he has seen that landmark, even though he disavows any tendency to progress in evolution. Arms races lead to bigger brains in mammalian herbivores and the carnivores that prey on them, notes Richard. "We seem to be seeing . . . an arms race, or rather a series of restarting arms races, between carnivores and herbivores," he writes. "This is a particularly pleasing parallel with human armament races, since the brain is the on-board computer used by both carnivores and herbivores, and electronics is probably the most rapidly advancing element in human weapons technology today."

The pure Spencerian view of the world, therefore, is that increased complexity is an inevitable manifestation of the system and is driven by the internal dynamics of complex systems: heterogeneity from homogeneity, order out of chaos. The pure Darwinian view is that complexity is built solely by natural selection, a blind, nondirectional force; and there is no inevitable rise in complexity. The new science of Complexity combines elements of both: internal and external forces apply, and increased complexity is to be expected as a fundamental property of complex dynamical systems. A fundamental property of complex adaptive systems is the counterintuitive crystallization of order—order for free, in Stu Kauffman's terms—upon which selection may act. Such systems may, through selection, bring themselves to the edge of chaos, a constant process of coevolution, a constant adaptation. Part of the lure of the edge of chaos is an optimization of computational ability, whether the system is a cellular automaton or a biological species evolving with others as part of a complex ecological community. At the edge of chaos, bigger brains are built.

Is human consciousness to be found there, too?

CHAPTER EIGHT

The Veil of Consciousness

"It was an extraordinary experience," recalled Chris Langton. "It's difficult to describe in any precise way, but it was like my brain switched to a new level of activity. Maybe it was triggered by the heat stroke." We were at Chris's house, midway between Santa Fe and Los Alamos, and he was telling me about an odd aspect of his recovery from a devastating hang-gliding accident. "When my face smashed into my knee in the crash, I shook up my brain real badly, damaged it in a diffuse way, nothing specific. Generalized trauma, I think it's called. When I recovered initially, I wasn't the same 'me'; I knew that very clearly. There was some of 'me' missing. Then, every once in a while I'd wake up and some part of 'me' would be back; like booting up a computer to a new level. It still plagues me that I'm not the person I was, and never will be."

The accident had been in the fall of 1975. A decade after his close encounter with death he had become obsessed with founding a new scientific endeavor, that of Artificial Life. The first international workshop on artificial life, held at Los Alamos National Laboratory in September 1987, was a *de facto* recognition that Chris had succeeded. It was while he was preparing the introductory chapter to the proceedings volume of the conference—the same chapter that so inspired Tom Ray—that he experienced another "booting up" of his brain, the one that was perhaps triggered by heat stroke.

"There was so much going on at the lab that I took myself up to

Tsankawi Mesa to write," said Chris. "It's quiet up there, with spectacular views across the Rio Grande Valley, and it's a good place to think." The mesa is part of the Pajarito Plateau, in the Jemez Mountains, where the smell of pine and juniper hangs in the clean air and clear streams run even in the dry months. Anasazi Indians lived there in simple settlements when the Chaco Canyon community was at its height. When Chaco collapsed in the late twelfth century, many Chacoans moved to this part of the Rio Grande Valley, where drought had not reached. "I must have spent too many days up there," Chris continued. "It's hot and the air is dry, and even though I took water with me, the sweat mechanism must have broken down, and I got heat stroke."

By the time he got back home at the end of the fourth consecutive day visiting the mesa, Chris was suffering an excruciating headache and a rapidly rising fever. In the middle of the night, with these symptoms growing alarmingly worse, he dragged himself to the hospital and had to be revived with a saline drip. "Eventually I went back home, and slept for a long time. When I woke I was aware that I'd got back something of the 'me' I'd been missing. It was a sense of my presence in the world." Before the return of this missing part of himself, Chris felt he was living in the middle of a cube, the sides of which were cinema screens with pictures projected on them. "It's hard to describe," he told me. "It was as if I could see the world, but somehow I wasn't in it, no emotional presence. Like looking at a picture of something rather than seeing the real thing and reacting to it as a person. I was aware of what I was missing, but I couldn't conjure it up. It distressed me a lot. Then it came back, just like that." Shortly afterward, Chris returned to Tsankawi Mesa, to see it again for the first time.

I tried to imagine viewing the world as Chris had for a while, but couldn't. I simply couldn't imagine away part of the thought processes that make "me" what "I" am. It sounds like an aspect of consciousness, I said. "Yes, I think it is," Chris answered thoughtfully. "And even though the earlier 'me' experienced the world like that, the new 'me' finds it difficult to recall clearly what it was like and still more difficult to convey to someone else."

When I had first questioned Chris about the scope of the new science of Complexity, about a year earlier than this conversation, I had asked whether the new science of Complexity could explain consciousness. "If the theory of complex systems is not some kind of seductive illusion; and if the brain can be described as a complex adaptive system; then, yes, consciousness can be explained, too," Chris had replied, confidently, and then qualified it: "At least in principle." I reminded Chris about this, and said, Do you really think that the kind of thing that happened in your head that day, the kind of sensation we all experience in our heads, is tractable to what Complexity has to offer? "Maybe not what Complexity *has* to offer, but what it *will have* to offer," Chris replied.

We were sitting at a round table in a dining room and kitchen that flowed into one space, with wooden floors, white walls, and beamed ceilings. "We're *serious* cooks," Chris had told me earlier. I could see the accoutrements of serious cooking everywhere, and heard of imminent plans for expanding the house, which would involve the installation of a professional stove at the center of a much enlarged kitchen. For now, things were a little cramped, and Chris had difficulty finding paper and pen with which to make his point. "I'm convinced consciousness is a bottom-up, emergent phenomenon," said Chris, beginning to sketch. He was drawing versions of his favorite diagram, which shows global properties arising from local interaction, the iconic image of emergence in complex systems.

"OK, so I think consciousness is probably five or six levels up," he said, drawing more diagrams. You mean you have a series of systems, each producing some kind of global property, and these global properties interact with each other to generate another level of emergent properties, and so on through five or six levels? "That's what I mean. It's hierarchical, many levels up, and you'd have extremely distributed properties." That looks horribly complicated I said, difficult to get a handle on. "I'm sure it is, and it may be impossible to describe the behavior at the highest level. We may need to know the behaviors of the parts at some of the lower levels."

The image was powerful, but elusive, too. Consciousness as an emergent phenomenon from a complex adaptive system? It sounded right, but I wondered how it might be instructive beyond mere description. And what of the other two pillars of complex adaptive systems: the crystallization of order, and complex computation at the edge of chaos? I needed to find out how, if at all, these might illuminate the phenomenon of consciousness. The science of Complexity had proved a powerful, if intransigent, tool; penetrating the veil of consciousness would be a tough challenge for it, perhaps the toughest of all. I knew that Jim Watson, codiscoverer of the structure of DNA, had recently described the human brain as "the most complex thing we have yet discovered in our universe." And consciousness may be the biggest puzzle that emerges from that two pounds of soggy gray matter.

The Princeton psychologist Julian Jaynes wrote: "Few questions have endured longer or traversed a more perplexing history than this, the problem of consciousness and its place in nature. . . . Something about it keeps returning, not taking a solution." To judge from the flourishing industry of book publishing on the subject in recent years, our thirst for finding out what this "something" is shows no sign of diminishing. And it is surely significant that the number of such books—by eminent scholars in philosophy, psychology, neurobiology, computer science, and other disciplines—is matched only by the diversity of their conclusions about the nature of consciousness and its generation in the human brain. We gulp down each offering, but the thirst remains unslaked.

To University of Washington neurobiologist William Calvin, for instance, consciousness consists of "contemplating the past and forecasting the future, planning what to do tomorrow, feeling dismay when seeing a tragedy unfold, and narrating our life story." For Cambridge University psychologist Nicholas Humphrey, an essential part of consciousness is "raw sensation." Roger Penrose, a mathematical physicist at Oxford University, suggests consciousness is "the ability to divine or intuit truth from falsity in appropriate circumstances—to form inspired judgements." According to Stevan

Harnad, editor of the respected journal *Behavioral and Brain Sciences*, "Consciousness is just the capacity to have experiences."

Each of us has a sense of what is meant by consciousness. I use the word "sense" advisedly, for whether we think of the processing that underlies consciousness as being mere computation or something more numinous, a strong feeling of self surely intrudes. That sense of self, which seems to exist as a separate entity from our physical self, is the source of wonder and mystery with which we contemplate consciousness. So it was for the French philosopher René Descartes, who, three and a half centuries ago, wrote: "So serious are the doubts into which I have been thrown . . . that I can neither put them out of my mind nor see any way of resolving them. It feels as if I have fallen unexpectedly into a deep whirlpool which tumbles me around so that I can neither stand on the bottom nor swim up to the top." Descartes's solution to the bewildering mystery of the mind-body problem, as it came to be known, was to say that the sense of self and the physical self were indeed separate, a philosophy known as dualism: the mind resides in the body, but is discrete from it.

Cartesian dualism dominated philosophical thinking for three centuries, until the British philosopher Gilbert Ryle effectively demolished it in his 1949 book, *The Concept of the Mind*, with the cutting phrase "the dogma of the ghost in the machine." True, Cartesian dualism is not completely dead, as evidenced in the views of Sir John Eccles, one of this century's greatest neurologists. In his *Evolution of the Brain*, published in 1989, he wrote: "Since materialist solutions fail to account for our uniqueness, I am constrained to attribute the uniqueness of the Self or Soul to a supernatural spiritual creation," which, he said, is "a miracle for ever beyond science." But for the most part, materialism, the philosophical alternative to dualism, dominates modern thinking about consciousness. As Tufts University philosopher Dan Dennett puts it: "The mind is somehow nothing but a physical phenomenon. In short, the mind is the brain." The debate these days, therefore, is among materialists, who argue over how consciousness arises from the physical stuff of the brain (although some would contend that dualism sneaks in here and there in different guises, not least of which is in some of the assumptions of artificial intelligence research).

I decided I needed to talk to Dan Dennett, who has a reputation as a wide-ranging and inventive thinker, and has contacts with the Santa Fe Institute folk. His basement office in the Eaton building of Tufts University is small, square, and windowless. A blackboard extends along one wall, completely clean. A voodoo mask, with two mouths, two noses and three eyes, stares down from a shelf. Below that is a white phrenology bust and a transparent plastic head filled with electronic circuits, icons of nineteenth- and twentieth- (perhaps twenty-first-) century views of the mind. A small table in the corner is a scatter of books (including Penrose's *The Emperor's New Mind*) and a copy of the *Times Literary Supplement*, in which Dan excoriates one of the recent crop of books on consciousness. Bearded and avuncular-looking, Dan has firm judgements and is quick to make them known. He picked up another of the recent books on consciousness and sniffed, "A disreputable piece of philosophy," and tossed it back on the table.

His new book is *Consciousness Explained*. That's an ambitious title, I ventured. "Yes," he acknowledged, laughing. "Actually I don't claim to have all the answers, or even most. But I do think I've made important progress toward what we're trying to explain." His message is twofold. First, that the notion that there is something within the brain that monitors sensations and thoughts, thus generating a conscious self, is wrong. Dan characterizes this concept as the Cartesian Theater. Second, that the sequential stream of consciousness we experience is an illusion, the filtered product of what he calls multiple drafts of mental states. Dan's is a cerebral view of consciousness, a focus on the higher levels of self rather than raw sensations, or what philosophers call qualia. It is also a view with which advocates of "strong" artificial intelligence can identify, that computation *is* mind.

In his book Dan wrote that the suggestion of some kind of monitor in the brain is "the most tenacious, bad idea bedeviling our attempts to think about consciousness." Strong words, I said.

"It's difficult to escape," replied Dan. "I am the observer of my consciousness, and you are of yours, but the bad idea is that there is an observer within the observer, what used to be thought of as a homunculus in the brain, watching what was going on, pulling levers, pressing buttons. It's a bad idea because we have to get away from thinking of a brain area putting out messages of the kind: 'There's a man in a blue suit approaching.' In fact, you have to think about decentralized, distributed systems, and that's difficult to do. It's one reason why the Cartesian Theater concept is so tenacious. The messages that brain areas put out are really very basic. Here, look at this." Dan handed me a Gary Larson cartoon that showed a man walking a dog, with lots of other dogs nearby. "Dog language translated," read the caption. And in voice balloons from each dog was the following: "hey hey! hey hey!"

"It's one of his most brilliant," said Dan, looking again at the picture as I handed it back to him. "You see, if we could listen in directly to what each of our brain areas was saying, it would be, 'hey hey! hey hey!' And out of this monosyllabic conversation from many brain areas, the whole system gets informed about the man in the blue suit." Instead of a single homunculus sitting at the center of the brain, there is a Pandemonium of Homunculi, says Dan, drawing on an image concerning an idea of architecture in artificial intelligence. "Information is flowing in from many senses, and it is subject to continual editorial revision, which produces multiple drafts of narrative fragments all over the brain."

That really does sound like pandemonium, I said. How do you get anything sensible out of it? "I want you to imagine something I call the Joycean Machine, which filters the multiple drafts and ultimately gives the illusion of a single, stream-of-consciousness narrative," Dan replied. "We're looking at the emergence of coherence from a massively parallel processing machine, the brain. You can think of it as a virtual machine. Counterintuitive, yes. Difficult to accept, true. Outrageous, I'll grant you that. But what would you expect from something that has to break through centuries of mystery, controversy, and confusion?" Dan is nothing if not bold.

The way you describe it, I said, resonates closely with the Santa

Fe Institute approach, doesn't it? "Absolutely right," Dan responded. "Emergence is what my model and their approach have in common. A few years back I had to talk about 'innocently emergent features' so that the biologists wouldn't get upset and think I was talking about something mystical. But, yes, emergence is a real, hard science phenomenon, and it's central to understanding consciousness."

Your model places a lot of emphasis on language, I said. "It does, and for good reason," said Dan. "Add language to the brain, and there's so much more you can do with the hardware. Without it the multiple drafts model couldn't work." So you are denying this kind of consciousness to all animals but humans? "I am. You know, Wittgenstein once said, 'If a lion could talk, we could not understand him.' I don't think that's correct. I think we would be able to understand the lion, but we wouldn't learn much about the life of ordinary lions from this talkative one, because language would have vastly transformed his mind."

No animal without language experiences a sense of self, argued Dan, not in the way that humans experience self. No multiple drafts, no stream of consciousness, just a biological self. "Can I prove that a bat doesn't have these mental states?" Dan asked rhetorically. "No I can't. But I also cannot *prove* that mushrooms aren't intergalactic spaceships spying on us."

Dan's book has been widely praised, his model said to be inventive and powerful, but he is criticized for aiming too high with what consciousness is. "People say I leave out qualia, but I think I address that," he said, referring to the more basic level of consciousness—the level concerned with simple sensation. A couple of years ago Cambridge University psychologist Nicholas Humphrey spent a sabbatical leave at Tufts, specifically to talk to Dan about consciousness. It was an intense, creative period, and the two men wrote a paper, called "Speaking for Our Selves," which examined consciousness from the point of view of multiple personality syndrome. They also talked a lot about qualia. "Dan's multiple draft model is excellent," Nick told me, "but there's no doubt in my mind he misses out on qualia, or raw feelings."

I've known Nick for twenty years, and I've watched as his own ideas of consciousness have evolved. During the 1970s he created a tremendous impact by asking and answering the question: What is consciousness *for*? His answer was that it had evolved as a device for playing social chess, the complex social interaction and manipulation that goes on in the lives of higher primates, and particularly in humans. An individual, by monitoring its own feelings and reactions to situations, is able to predict more accurately the reactions of others, thereby gaining an advantage in the game of social chess. Nick's notion of the social function of intellect and consciousness became, and still is, a favored explanation among anthropologists and primatologists for the evolution of unique features in the higher primate brain. For instance, a recent major review of primate cognition said that "among nonhuman primates, sophisticated cognitive abilities are most evident during social interactions with [other members of the troop]."

"This view of consciousness, like Dan's, focussed on higher-level, second-order consciousness," said Nick. "I still believe that self-awareness, which humans experience and chimpanzees do, too, but to a lesser degree, is important in the social context. It allows us to model our own minds, and it was a crucial factor in becoming human. But I grew more and more uncomfortable with it, and now I have a different view of what consciousness is." Nick has contributed to the consciousness book industry, with *A History of the Mind*. Its thesis is that consciousness is sensation, raw feeling, no more, no less: sensations of color, sensations of pain, sensations of hunger—un-thought-out, uncategorized, prepropositional experience. "Feelings enter consciousness, not as events that happen to *us* but as *activities* that we ourselves engender and participate in—activities that loop back on themselves to create the thick moment of the subjective present," he wrote.

That sounds pretty basic, I said. "It is," Nick replied. And you're not abandoning the social function of consciousness in humans? "No, I'm not. Humans experience that, no question about it, and we know it as introspection. It's very important in the way we conduct our lives and how we feel about our lives. But by restricting consciousness only to the cerebral level as I had, I excluded most of the rest of the animal kingdom. By arguing as I do now that sensation in the present constitutes consciousness I can bring a lot of animals back into the fold." I asked why he wanted to do that; it clearly was not an easy question to answer. "Well, I felt in my bones it wasn't true," he began, defensively. "What I had been talking about was the ability to reflect upon a state of mind, not simply the state of mind itself. The more I argued that case the more blank looks I got, and the more I began to sympathize with those blank looks."

You're saying that you now consider that the state of mind itself constitutes consciousness, and that the ability to reflect on it is something extra, a secondary level of consciousness that only humans experience? "Yes, that's where I've come to," said Nick. "The more you look at animals the more difficult it becomes to deny them sensation. Animals at some level know they have pain; they're just not aware of it in the way we are. It's the present that's crucial in consciousness, not reflecting on the past or the future." By extending consciousness way out into the rest of the animal kingdom, you are making humans seem somewhat less special, aren't you? "Yes I am, and I think there's a strong urge gathering in that direction," said Nick. "People seem to want to believe in some kind of continuity between us and other animals. That's not to deny humans have special qualities. We do, but we also share this basic level of consciousness with them. I think the urge for reestablishing continuity is a reaction to the arrogance of the artificial intelligence people who claim to have solved the problem of consciousness at a mechanical level."

I asked whether he thought a computer would have no facility for feelings, even if it were running programs that mimicked human thought. "Thinking machines are not difficult to build," responded Nick, "but they are not feeling machines." Today's supercomputers, particularly massively parallel computers that are a step closer to brain architecture than conventional serial computers, are able to achieve respectably powerful thought processes. But that's not enough to engender feelings, argues Nick. "The reason computers can't feel is that they have nowhere to feel anything. Computers come in a box, and that's not a significant boundary for it. The box we come in is the boundary of our experience, and our sensations are the experience of what is happening at that boundary."

Two theses about consciousness; in fact, two very different forms of consciousness. Dan Dennett's is a higher-level, computational phenomenon. Nick Humphrey's is a basic, essentially noncomputational sensation. Whenever philosophers or psychologists talk about consciousness they are always aware that looking over their shoulders is the artificial intelligence community. A mix of philosophically inclined thinkers and inspired doers, the community approaches the human mind in a way best summed up by Marvin Minsky's description of the human brain: "a computer made of meat."

To say that the brain is a computer is a truism, because, unquestionably, what goes on in there is computation. But so far, no man-made computer matches the human brain, either in capacity or design. Danny Hillis, the scientific inspiration behind the world's most advanced computer, the Connection Machine-5, describes his machine as "trivial in complexity compared with the brain of a fly." Nevertheless, the question can still be asked, Can a computer think? And, ultimately, can a computer generate a level of consciousness that Dan Dennett or Nick Humphrey, or anyone else, has in mind?

Famous in the science of artificial intelligence is the Turing test, a Rubicon that separates mere computing from mindlike computation. Formulated in 1950 by British mathematician Alan Turing, the challenge is to create a computer system that can fool an interrogator into thinking that he or she is having a dialogue with another human being, not a machine. To advocates of strong AI, a computer that passes such a test is not merely a model of the human mind; it *is* human mind in a very real sense. According to this view, mind—that is, cognition and consciousness—results from running the right program, no matter whether the hardware is formed from silicon or lipid membranes.

No computer has passed the test in any convincing way so far, though some limited successes have been scored recently. Even if a computer should pass the test, there are many who will remain unimpressed, but for different reasons. "The Turing test enshrines the temptation to think that if something behaves as if it had certain mental processes, then it must actually have those processes," John Searle wrote recently in an article in *Scientific American*. For example, the fact that the computer system Deep Thought can compete at the grand master level of chess says nothing about the system's grasp of the game. The computer attains its competitive level by its ability to run through seven hundred thousand possible moves each second, not through creative strategy. Deep Thought plays chess well, but it is not a chess player.

Searle, a philosopher at the University of California, Berkeley, suggests that the advocates of artificial intelligence are unwittingly pursuing a new form of dualism. "Unless one accepts the idea that the mind is completely independent of the brain or of any other physically specific system," he wrote, "one could not possibly hope to create minds just by designing programs." Searle believes such a quest to be futile.

In the same issue of *Scientific American*, Paul Churchland and Patricia Churchland, philosophers at the University of California in San Diego, also reject the Turing test as inadequate for recognizing minds, but for different reasons. Like Searle they argue that what goes on inside the computer is an important criterion of mind, but they allow for the possibility that one day a mindlike computer could be built. This would require a shift from conventional serial-processing to parallel-processing machines. "Artificial intelligence, in a nonbiological but massively parallel machine, remains a compelling and discernible prospect," they wrote.

The most vigorous and public attack on the computation-equals-mind school of artificial intelligence was Roger Penrose's recent book, *The Emperor's New Mind*. "I thought of the title before I wrote the book," Roger told me when we met in his office at the Mathematics Institute, a modern building in St. Giles, surrounded by Oxford's ancient colleges. "Then I found that few people understood what I meant by the title, so I had to write a little story to wrap around the book."

The story, set in the future, is of the unveiling of a computer that is so powerful it would effectively run the affairs of state. Packed with 10^{17} "logical units," the machine would surely outstrip any human brain or committee of them. At the moment of unveiling, the chief designer asks if anyone would like to pose a question to this ultimate mind, by way of initiation. A small boy stands up, and asks, "What does it feel like?" Much derision is expressed at so naive a question, and the chief designer reports that the computer doesn't understand what the boy means. "In the French translation the point was missed completely, and the publisher wanted to change the title," Roger said. "The point is that the advocates of AI really are deluding themselves and others when they argue that computation *is* mind." Are you saying that no mental activities are computational? I asked. "No; some surely are. But when we're talking about consciousness and creativity, the computational analogy is inadequate. Algorithms are inadequate as a means of achieving consciousness and creative thought."

I asked whether the new computational possibilities opened up by massively parallel processing computers might approach creativity and consciousness. "No, I don't think so," Roger replied. "You're still talking about algorithms within the realm of mathematics as we know it, and what I'm looking for is something outside that." Isn't that verging on the mystical? I asked. "It can seem like it, and, I have to admit, I sometimes feel sympathetic to mystical interpretations. But, no, I'm looking for a new quantum physics of the mind."

The Emperor's New Mind is a *tour de force* of mathematics, physics, and philosophy, one long argument about the inadequacy of the computational model of consciousness. Nevertheless, one reason why Roger plunged in "where mathematical physicists probably shouldn't venture" stemmed from personal experience. Important ideas have frequently come to him fully formed, leaping out of a cognitive foment and requiring only to be tidied up. "It seems to me that whenever I have a mathematical idea my mind is making contact with Plato's world of mathematical concepts," Roger explained. "It's like reaching into Plato's ideal world and retrieving something that already exists. This does not feel like an algorithmic process of discovery. It feels like something quite different, and as yet, in my opinion, it cannot be explained by anything that the artificial intelligence people talk about. It is not mere computation." I felt, once again, I was hearing the drumbeat of emergence here, distant but distinct.

Neuroscience is said to be awash with data about what the brain does, but virtually devoid of theories about how it works. Some overall descriptions of the properties of the human brain are instructive. For instance, 10 billion neurons are packed into the brain, each of which, on average, has a thousand links with other neurons, resulting in more than sixty thousand miles of wiring. Connectivity on that scale is beyond comprehension, but undoubtedly it is fundamental to the brain's ability to generate cognition. Although individual events in an electronic computer happen a million times faster than in the brain, its massive connectivity and simultaneous mode of activity allows biology to outstrip technology for speed. For instance, the fastest computer clocks up a billion or so operations a second, which pales to insignificance beside the 100 billion operations that occur in the brain of a fly at rest.

The magic of it all is that while no single neuron is conscious, the human brain as a whole is, and it generates the leaps of creative insight that so impress Roger Penrose and others. How does it do it? How are simple electrical signals across individual cell membranes transformed into cascades of cognition? How are billions of individual neurons assembled into a brain, seat of the mind?

Patricia Churchland, a philosopher, decided some years ago that if she was to understand how the mind works, she would need to know some neurobiology. Neither philosophy on its own nor neurobiology on its own could promise an answer, she believed. Her 1986 book, *Neurophilosophy,* was the first stepping stone in the link between the two disciplines. Her 1992 book, *The Computational Brain*, coauthored with Terrence Sejnowski, is the second. We met at a scientific conference in Berlin, and after a group visit to a concert at the Philharmonic Hall we headed for a local restaurant to talk about consciousness. Appropriately enough, it was the Cafe Einstein.

Is it reasonable to think of the human brain as a complex dynamical system? I asked. "It's obviously true," she replied quickly. "But so what? Then what is your research program?" Pat combines an appreciation of brain function as a whole with a scrutiny of the mechanics of individual systems within it. "Nature is not an intelligent engineer," she continued. "It doesn't start from scratch each time it wants to build a new system, but has to work with what's already there." That's François Jacob's notion of evolution as *bricolage*, cobbling together contraptions from whatever is available, isn't it? "That's right, and the result is a system no human engineer would ever design, but it is wonderfully powerful, energy efficient, and computationally brilliant. It is also a marvel of miniaturization. Nervous systems evolved, and that makes it difficult for neurobiologists, and especially AI people, to look at the wiring diagram and figure out what's going on." Why especially artificial intelligence people? I asked. "Because they tend to approach the problem within the framework of electrical engineering, and with prejudices about how they think brains *should* process information, instead of finding out how they *do*."

There is an increasing trend among artificial intelligence researchers to move away from conventional serial-processing computers to parallel-processing machines, which Pat applauds. "It's obviously the direction to take," she says. "The nervous system is a parallel-processing device, and this conveys several interesting properties. For a start, signals are processed in many different networks simultaneously. Next, neurons are themselves very complex little analogue computers. Last, the interactions *between* neurons are nonlinear and modifiable. Real neural networks are nonlinear dynamical systems, and hence new properties can emerge at the network level." You can therefore describe the output as an emergent property, can't you? "That's right, you can. When you think about brain activity it's correct to think about emergent properties at higher levels that depend on lower-level phenomena in the system."

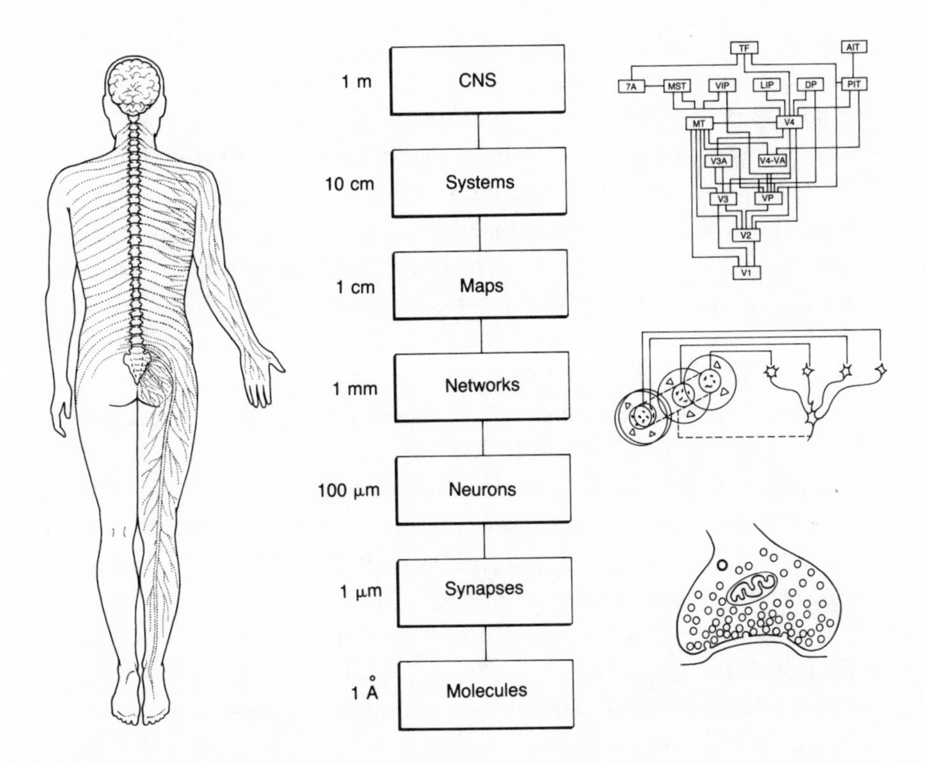

Fig. 8. Structural levels in the organization of the nervous system, a reflection of the hierarchical systems that may underlie the generation of higher cognitive functions, including consciousness. Courtesy of Patricia Churchland and Terrence Sejnowski.

I explained about the tenets of the new science of Complexity, the importance of emergence from dynamical systems, the counter-intuitive notion of the crystallization of order from complex networks, the computational power at the edge of chaos. "Stuart Kauffman's idea of innate order in networks has the right kind of feel about it for some aspects of brain operation," Pat responded. "But again, you're faced with the question: Then what research do you do? I prefer to take the route through the more basic level. Theories have to be testable, and testability is more feasible that way."

And what of the edge of chaos notion? I asked. "Yes, there could be something in that. It might give a framework for us to come to grips with some higher function puzzles. But before we address the neurobiology of creativity and unpredictability, we need to understand the precision and predictability in the nervous system. How does an owl succeed so often in catching the scurrying mouse? How does a flying bat succeed *so often* in intercepting a flying moth? How do I generally manage to *say* what I intend? These feats of nervous systems suggest there is tremendous precision, adaptability, and predictability. My personal hunch is that the hideously complex problems can best be approached after we have in hand solutions to less complex problems—as in physics, where you get hopelessly stuck if you insist on understanding turbulence before you understand how balls roll down an inclined plane."

I, too, felt that, at least on the level of analogy, the edge of chaos was rich in meaning: a system poised to respond, nudged into creative activity by simple perturbations. But, yes, how does one go from fruitful analogy to experiments at the bench?

"I'd like to go back to something I was saying about networks," said Pat. "It will give you some idea of what we're up against." She described a network of neurons now famous among neurobiologists, known as the stomatogastric ganglion in the spiny lobster. The ganglion, which contains about twenty-eight neurons, drives the rhythmic muscular motion of the animal's gastric mill. Allen Selverston, of the University of California, San Diego, has performed a heroic study of the ganglion. "A tremendous amount is known about its overall anatomy, its network connections," Pat explained. "We know which neurons talk to which others, and with what effect. But, even with all this information, a very rich description of the network, we still don't understand how it produces the rhythmic output we see. The message is that the details of the system are necessary if we are to understand its activity, but they are clearly not sufficient."

An analogy occurred to me. You know that, even given the complete DNA sequence of an organism, molecular biologists cannot deduce how that organism assembles itself during development, I said. Something more is needed. (I was thinking about Brian Goodwin's holistic approach to development.) So, in the same way, even if you had the complete wiring diagram of the human brain, you still wouldn't be able to say how cognitive processes arose from it? "You're right. It's not enough to know the micro-architecture. We also have to understand the network properties that arise from the micro-architecture, and so far that's not at all obvious." That, Pat said, is the message of her new book.

Shifting intellectual gears, we talked about the genuinely mysterious qualities that consciousness subjectively possesses. Even an understanding of the phenomenon might not remove that, we agreed, and in any case, *having* an experience of blue is completely different from *knowing* the brain mechanisms for the experience. Then, in speculative vein, Pat said: "We do our research as if materialism was a proven fact, but of course it isn't." You mean, Cartesian dualism could be true? "I mean, we cannot claim to have ruled it out. The mind-body problem has been a mystery for so long that you can understand the appeal of the idea that there really is something else beyond what we know about the physics and chemistry of the brain, or even what we *can* know. I do not for a minute think there might really be a nonphysical soul, but I also realize we have a lot to learn about how the brain works."

Colin McGinn, a British philosopher at Rutgers University, is among the most articulate scholars currently addressing the mysterious nature of consciousness. He has an unusual take on the issue, however. "The mystery is real," he told me, "but I'm not arguing that there's something magical about consciousness. I'm as much of a materialist as anyone. What I argue is that an understanding of consciousness is beyond the reach of the human mind, that cognitively we are not equipped to understand it in the way we understand other phenomena we experience in the physical world."

Colin's argument is rooted in what he calls biological realism. Simply put, it is this: just as the brain of an oyster is limited in what it can encompass, so too is that of a rat, a monkey, and a human. "Complete cognitive openness is not guaranteed for human beings and it should not be expected," he told me. "The deep feeling of mystery we experience with respect to consciousness should at least encourage us to explore the possibility that an understanding of it is simply closed to us." Human senses are geared to representing a spatial world, Colin explained. Because consciousness fundamentally is subjective experience, our analytical senses, which are so successful in exploring the rest of the natural world, simply fail to encompass it.

"You can analyze brain structure and function in the way we analyze other phenomena," said Colin, "but the information you get tells you about nerve cells and circuits. Alternatively, you can think about consciousness as subjective experience. And what you find is that the two sides of inquiry never meet and, I think, never will." I asked Colin whether he was saying that consciousness was generated outside the physics and chemistry that we know, something akin to Roger Penrose's argument. "No, as I said, I'm as much of a materialist as anyone. There's nothing mysterious about the physics and chemistry underlying consciousness. Our problem is that the phenomenon that arises from that chemistry and physics—consciousness—isn't available to the kind of analytical thinking of which humans are capable."

Dan Dennett is deeply scornful of this line of argument. "I think it's objectionable," he told me. "It's framed in a pseudo-biological way, saying that the oyster and the ant have limitations, and so must we. Language so transforms our minds that we are on a different scale." The motivation for Colin's line of argument, Dan ventures, is "to build a Maginot Line around the mind so that scientists can't get at it." His indignation was barely containable. "It's religious doctrine," he snorted finally.

"Dan's position is a massive piece of dogmatism," Colin told me. "My argument is the strongest form of naturalism you can imagine. What I'm saying, and what Chomsky has said for a long time, is that we have to be naturalistic first about our own cognitive abilities." Colin argues that we find it easy to accept that, unlike some other creatures, humans can't see ultraviolet light, for simple biological reasons. Humans can't hear ultrasonic sound, while some other creatures can, again for simple biological reasons. So why should humans expect to be able to understand every phenomenon that emerges from the brain? "When I say consciousness is a mystery, I'm making a naturalistic point about human cognitive abilities, not about any mystical quality of consciousness itself. Consciousness may be a rather simple biological characteristic, like digestion."

"Of course consciousness *seems* mysterious, but that's just the subjective element that we humans experience," said Norman Packard. We were talking in the Santa Fe offices of the Prediction Company, with secret algorithms on powerful computers exploring the mysteries of financial markets in several side rooms as we sat in a light-filled room at the back of the building. "But I don't think it's a mystery in any important sense, in our urge to try to understand it."

By now I had formed an image of consciousness as the most far-reaching intellectual challenge facing the new science of Complexity, a phenomenon of mercurial properties. William James, in his *Principles of Psychology* (1890), wrote: "As we take . . . a general view of the wonderful stream of our consciousness, what strikes us first is the different pace of its parts. Like a bird's life, it seems to be made of an alternation of flights and perchings." In their quest to understand consciousness, modern scholars apparently cannot agree on which direction the bird is flying, where it might perch, nor even what nature of bird it is. A mystery indeed.

I wanted, finally, to discover what the science of Complexity might bring uniquely to this quest. I had heard that at an early meeting of the Santa Fe Institute, at which the scope of its investigations were explored, people shrank away from the challenge of consciousness. Philip Anderson had repeatedly taunted the gathering by asking, "What about the 'C' word?" He had no takers. It seemed to me, after my odyssey through all the patterns of nature, that one promising line beckoned in this context: that of the drive of complex adaptive systems toward information processing. I also remembered that at the 1987 meeting on Evolutionary Progress at Chicago's Field Museum, Francisco Ayala had described human consciousness as "the climax of one kind of progress, that of information processing."

I asked Norman whether he thought this intuition was valid. "Absolutely," he replied. "The idea is a natural. In the evolution of the biosphere you see computation and information processing happening at different levels and different places. You have information processing within organisms, within cells of organisms, and within units comprised of many organisms." You mean, as in ant colonies? "Yes, and in colonies of other social insects. And of course in human society."

What kind of information are we talking about here? I asked. "Raw sense data, and these get processed into some kind of representation of the world." But this surely doesn't have to rise to the level of awareness for the organism to be able to operate, does it? Organisms could process this kind of information as efficient automata. "That's right, but don't you think that your impression of the world, through self-awareness, influences how you think other animals experience their worlds? I think they have a level of consciousness that isn't necessarily as sharp as ours has become, because of this extra phenomenon of self-awareness. Consciousness isn't a binary phenomenon, on or off. There are degrees of it."

OK, I said, can the science of Complexity bring anything unique to the study of consciousness? "Ultimately, yes. The way I see the science is that it's concerned with information processing throughout the entire biosphere; information processing is central to the way the biosphere evolves and operates. Consciousness is just one part of that larger puzzle, and it's important to remember that. Most studies of consciousness focus just on the phenomenon itself, and that's solipsistic. I'm not saying that's invalid, but you asked what unique contribution the science of Complexity could bring to the endeavor, and that is to place consciousness into the larger puzzle of information processing in the biosphere."

I have to admit to being unprepared for the forceful line of arguing I got from Norman. I had brought up the topic earlier with Chris Langton, Stu Kauffman, and others, and had discerned arguments that, in principle, the science of Complexity must somehow be able to address consciousness, but little more. Norman seemed prepared to go beyond that. In a quiet, determined mode of speech, frequently punctuated by long, thoughtful pauses, Norman gives the impression of seeing through a window into the future, a view not available to most.

That sounds impressive, I said, but can you actually bring it to earth in any practical way? "Oh, I think so," Norman responded. "The simple evolutionary models of the sort I'm working with will eventually develop behavior rich enough that I'll see some kind of consciousness emerge." You're saying that your computer model, a form of artificial life, will develop consciousness? "I'm saying that the level of information processing in the system will evolve toward what we could call consciousness, that the organisms will reach a point where they will do information processing on their own, and become aware." Artificial life, becoming aware of itself? "Yes."

Norman describes his computer system currently as extremely simple, having started with "stupid" organisms that "staggered around hardly able to find food." As the program evolved, however, the organisms improved, becoming more efficient at foraging for food, and even engaging in sex. "Sex is about as complicated an interaction as the organisms have with each other so far," Norman said. "But that's a start. There's no question in my mind that the improvement in their behaviors I've seen through evolution represents enhanced information processing strategies. One day they will evolve a kind of consciousness, I'm certain of that." But how would you know? I asked. There was a longer pause than usual. "Their brains are simple, and their world is different from mine, so, I don't know, it will be difficult."

Another pause. "If it comes to that, I know I'm conscious, but I don't know that you are."

CHAPTER NINE

The View from the Edge

According to the cover of the 22 July 1990 issue of the *New York Times* Sunday magazine, Edward O. Wilson is "The Ant Man." For good reason. Even before he reached his teens, growing up in a Baptist household in Alabama, Ed was a devoted naturalist and liked nothing more than puttering around in streams and woods. He never grew out of the bug phase through which many children pass, and now, half a century later, he is the Frank B. Baird, Jr., Professor of Science at Harvard and curator of entomology in the university's Museum of Comparative Zoology. Ants are everywhere in his large, square office on the fourth floor of the modern annex to the museum.

A car's license plate, from Georgia, reading HI ANTS, hangs on one wall. "A friend's," Ed explained. There's a giant picture of an ant on the refrigerator door. A bronze ant sculpture stands on a table in the middle of the room. A copy of *The Ant*, a 732-page compendium written with his colleague Bert Hölldobler, is displayed on a side table. Intended as a guide for would-be ant scientists, the wonderfully illustrated volume is so compelling in its descriptive prose that it won the authors the 1991 Pulitzer Prize (Ed's second). A box containing the computer game SimAnt, based on Ed's insights into ant life, is propped against a computer. And then there's the real thing, three colonies of leaf-cutter ants on tables on two sides of the room. Each colony is divided into two main parts, each with many compartments, and joined by a bamboo arch across which worker ants carry oat flakes (substitute leaves), which will be fodder for the colony's fungus garden.

"Wonderful creatures, aren't they," said Ed, as we watched the constant motion of the individuals in one of the colonies, countless fragments of activity melded to one purpose: the life of the colony. I asked whether these colonies had come from Finca El Bejuco, Tom Ray's patch of rain forest in Costa Rica. "Not these, but my earlier colonies did. These came from La Selva, nearby. But Bert and I were down at Tom's place a couple of years ago, collecting. So if you saw logs ripped apart along his trails when you visited, that was us." A tall, gangling figure, Ed watched the colony in silence for a few moments, absorbed. "We humans have a distorted view of the world," he said at length. "When we think about nature we usually think about creatures like us, large vertebrates. But vertebrates are rarities in the world of nature, compared with insects." And ants are king of the insects, or at least king of the jungle. A Smithsonian Institution scientist recently demonstrated that in the tropical forest canopy, ants make up 70 percent of the total insect population.

"You can think of ants as the culmination of insect evolution in the same sense that humans are the culmination of vertebrate evolution," Ed continued. "They both developed complex social systems, and that had a tremendous impact on their evolutionary success. Only 2 percent of insect species are social, but they represent more than half the insect biomass. And we can measure human success in our exploding numbers and the fact that we have colonized virtually every part of the globe. In fact, I'd say we're *too* successful." He did point out, with more than a glint of myrmecological triumph, that ants learned the trick of sociality a good 100 million years earlier than humans arrived on the scene. One up for the ants.

I knew early in my exploration of the biological implications of the new science of Complexity that at some point I would need to talk with Ed Wilson about ants. Ed is more famous—and, to some, infamous—these days as the "father of sociobiology," through the massive tome he published in 1975, titled simply *Sociobiology*. In it he argued that much of behavior, including much of human behavior, would eventually be understood in terms of genetic determination, a notion that some considered bold while others denounced it as fascist. Ed is fascinating, and convincing, on the subject, but it was ants that I went to see him about this time. There were two reasons, tightly linked. First was the biological impact of sociality, something that humans share with the social

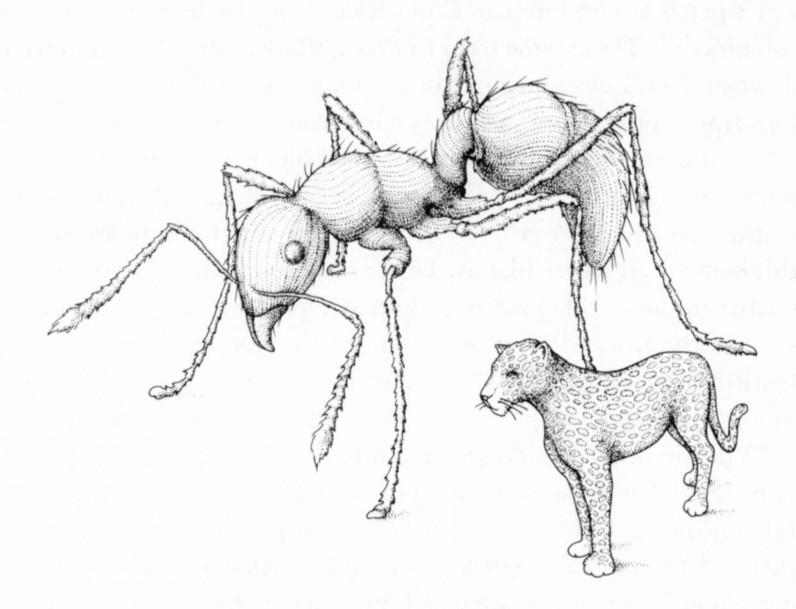

Fig. 9. In the Brazilian tropical forest, the biomass of ants is approximately four times greater than the biomass of all the vertebrates (mammals, birds, reptiles, and amphibians) combined, as shown by the relative sizes of an ant, *Gnamptogenys pleurodon*, and a jaguar. Courtesy of E. O. Wilson and Katherine Brown-Wing.

insects, particularly ants and termites. For the insects, sociality has become part of their inner nature, encrypted in their genes. For humans, sociality—at the level of complexity seen in insect colonies—emerged as a cultural expression of an inner potential, a much more dynamic property. Nevertheless, the commonalities are evident, and they are linked by the second of the two reasons: the phenomenon of emergence. The lives of individual ants and individual humans are transformed by membership in a larger entity, an entity they also help create.

Emergence, it had become ever more clear to me, is the central feature of the new science of Complexity. We saw it in the evolutionary models of Tom Ray and Kristen Lindgren, for instance, and in Stu Kauffman's models of coevolving systems. We saw it in the unfolding of morphological form in embryological development. We saw it in the properties of ecosystems, such as the existence of foodweb structures and the persistence of communities; and all the way up to global control, in Gaia. We saw it in the different levels of dynamic complexity in human societies, from bands up through to the state. And, at the level of detail at which Ed Wilson works, we can see it in the lives of social insects.

"Social insects pushed solitary insects to a minor position in the ecosystem," Ed explained. "The emergent properties of social life are so very powerful." There is something qualitatively novel about insect sociality, isn't there? "Certainly," replied Ed. "For a start, the colony as a whole processes more information," Ed replied. "An individual social insect processes less information than an individual solitary insect, but as part of an aggregate activity, the social insect contributes to more complex computation. The colony works as a single organism."

Early on in his studies, Ed realized the importance of communication in the workings of the colony. Much of the communication is in the chemical realm, as Ed and others discovered. In fire ants, for example, the nutritional needs of the colony are "known" by the whole colony, because the workers constantly exchange samples of their stomach contents, effectively creating a single stomach for the colony. Workers on the front lines of foraging therefore know what is going into the mouths of the young deep within the colony. "The mass response to requirements of the colony can be more precise this way than if each individual forager tried to assess the colony's needs itself," Ed explained.

One of Ed's favorite examples of colony communication was discovered by his colleague Bert Hölldobler. Honeypot ants, which live in Arizona, feed on termites when they can, a rich, abundant food source. Sometimes, however, scouts from two separate ant colonies encounter a termite colony simultaneously, and the right to the treasured resource has to be settled. Instead of an outright battle, the two ant colonies engage in a tournament in which squads of workers from each side strut around as though on stilts, and jerk their bodies as they approach an opponent. The display is confined to pairs or small groups in the opposing squads, which number up to two hundred individuals, and so represent just a small sample of the colonies. Rarely is there recourse to mandibular nipping or formic acid spraying, which make these creatures potentially deadly assault machines. Hölldobler discovered that typically the ant colony with the larger number of displaying foragers prevailed, winning access to the termites, "with little shedding of hemolymph," as Ed puts it. The colony makes the decision, the result of aggregate individual behavior.

"These examples are about as striking a demonstration of emergence as you could hope for," said Ed. "They give you some idea of why sociality is so successful in evolutionary terms." I knew that this success had been expressed many times in insect evolution. "Twelve times, in independent lineages," Ed informed me. "You can think of sociality as a biological attractor. It works with insects and with humans, but there's nothing with the same intensity of sociality in between." The phrase "biological attractor" was precisely how Brian Goodwin had described the production of biological form, including individual organs and whole organisms. Here, the phrase Ed applied to what organisms did, their collective behavior. "Obviously, with humans things are a little more complicated," Ed continued, "but human sociality is just as much of a biological attractor as it is in insects." You do see different levels of complexity in the sociality of different species of insect, but you don't see the progression through different levels—as human societies may progress through band, tribe, chiefdom, and state—within the same species. "Human sociality is more dynamic as a system," Ed observed.

There was something extremely pleasing in going from human sociality, which I touched on in my first contact with Complexity, and insect sociality, in this, one of my last interviews. With it, an intellectual circle closed, encompassing what would be a strong image of what the new science of Complexity might mean in the world of nature. For me, the process of exploration was nearing a close.

But I wanted to talk to Ed more about his reference to the ant colony—any social insect colony—as operating like a single organism. Four decades ago it was fashionable to refer to social insect colonies as superorganisms, and not as a mere analogy. To William Morton Wheeler, for example, Ed Wilson's predecessor at the Harvard Museum, an ant colony *was* a single organism: it displayed specialization of functions, the individual units were completely dependent upon the whole, which in turn was a consequence of their collective activity; and the end result was like nothing in the world of solitary insects.

"The superorganism was a beguiling idea, nice to talk about for a few minutes," Ed told me. "But it quickly wears thin, at least in the way it was viewed then, which, frankly, was rather mystically. Emergence was big at the time, too, but again it was heavily mystical." But you've talked about emergent properties in your ant colonies, I said. You weren't being mystical then, were you? "No, I wasn't. When I came here in the 1950s I pushed hard to get away from the superorganism concept, and wanted to ground our approach in obtaining details at a lower level." You would describe that as a reductionist approach? "Yes, I would. We needed to understand how parts of the system work before we could look at the whole. But now it's time to look at the whole once again and, yes, I think we can begin talking about insect colonies as superorganisms, but without the mysticism." You're not saying that knowledge of the lower-level details of how colonies operate is sufficient to understand the whole, are you? "No, I'm not. I'm saying there *is* something genuinely emergent about the behavior of a complex system like an insect colony, but that it's important in our understanding of it also to be acquainted with the mechanics of the system."

I explained that the concept of emergence was a vital part of the new science of Complexity, specifically in complex adaptive systems, whether they were in the realms of biology or physics. I also said that this presents a problem, as far as I could see, because modern biologists are suspicious of emergence as an explanatory concept. "Yes, many are, and for good reason," Ed replied. "By itself, emergence can be no explanation at all if you don't have any insight into the mechanics of the system, and it may seem to be an appeal to mysticism." But as a phenomenon, you're saying that emergence in biological systems is real. "Yes I am. There's no question about it."

By focussing on emergence as a biologically important phenomenon, the new science of Complexity has stumbled into a debate that has a long history and raw emotional content. For two millennia, an intellectual divide separated scholars' views of the natural world, one essentially Platonic, the other Aristotelian. On the Aristotelian side, mechanists said that living organisms are "nothing but machines," and are completely explicable by the laws of mechanics, physics, and chemistry. Platonics agreed that living organisms obeyed these physical laws, but insisted that the essence of life itself was something extra, a vital force breathed into mere material. To vitalists, therefore, many of the more interesting properties of organisms were, by their nature, beyond scientific analysis.

By the early decades of this century, the mechanists had prevailed, for two reasons. First, because scientific discovery had shown repeatedly that properties of organisms that previously were considered inexplicable indeed had mechanistic explanations. And second, mechanists had moved away from the strict "nothing but machines" position to accepting that living and nonliving objects were indeed different. The differences resided in the organization of physical material, so that organisms possessed properties not shared by nonliving objects. Mainstream biology therefore became essentially mechanistic.

The mechanists' victory was, however, never complete, with some philosophers and even some physicists explicitly promoting a form of vitalism. For example, in 1932, Niels Bohr, discoverer of the basic structure of the atom, said this: "The recognition of the essential importance of fundamentally atomistic features in the functions of living organisms is by no means sufficient for a comprehensive explanation of biological phenomena." Bohr's vitalism, which derived from his quantum physics, gained some popularity for a while.

At the same time, some biologists continued to argue that the laws of chemistry and physics alone were insufficient to explain important features of life, not because of the addition of some kind of *élan vital*, but because of emergent complexity. In 1961, Conrad Waddington put it this way: "Vitalism amounted to the assertion that living things do not behave as though they were nothing but mechanisms constructed of mere material components; but this presupposes that one knows what mere material components are and what kind of mechanisms they can be built into." Waddington was an emergentist, but not a vitalist. He believed that the assembly of a living organism is subject to physical laws, but that their product is not derivable from the laws themselves. In many ways, the new science of Complexity is heir to this line of reasoning. It is a new emergentism, a potentially far more powerful brand than any of its predecessors.

Nevertheless, advocates of Complexity are likely to find their message viewed with even deeper suspicion than any of its predecessors, principally because of the tremendous success of molecular biology during the past three decades, and particularly in the last. The tools for manipulating genetic material these days verge on the fantasy of science fiction, and the promises of even greater accomplishments are likely to be achieved. Organisms' DNA can be scrutinized in the smallest detail, the tiniest fragment of every message it encodes understood. Or so it is assumed. Modern molecular biology is therefore the ultimate in the reductionist approach to understanding organisms and their history, and represents the polar opposite of emergentism.

Not long ago I attended a small gathering of eminent scientists of different disciplines who were giving their views of the future of science. A Nobel Prize–winning molecular biologist stood up and said, "With our new ability to manipulate and analyze DNA, we can now begin to understand the process of evolution." He was serious. Simply read the messages in the genes, and all would be revealed—that was his view. No nod in the direction of the complexities of development. No indication that population biology may play a role in the fate of a species. No suggestion that species are part of ecosystems, which themselves are components of larger structures, all of which influences the unfolding of evolutionary history. And, of course, nothing at all about the immanent creativity of complex dynamical systems. As long as other biologists view molecular biology as the exemplar of modern biology—as many do—the phenomenon of emergence is unlikely to be instantly embraced as a powerful new insight.

Or is it?

"I believe we are on the cusp of an important change," Brian Goodwin told me. "The reductionism of molecular biology has been important, and no doubt we'll learn a lot more from it. But in the enthusiasm for gathering more and more data at what people view as the fundamental level of biological systems, the organism has been ignored. It's time for a change." Brian, one of Waddington's last students, continues his mentor's brand of mechanism in what some observers view as an odd blend of tough mathematics and Eastern mysticism. He is a theoretical biologist of the highest caliber, and yet, to the disquiet of some of his colleagues, often slips into deeply philosophical vein. "I completely reject true vitalism," Brian told me. "But, by taking the organism seriously in biology, by saying that there is some kind of organization that is distinctive to the living thing, we can move to a closer appreciation of the quality of the organism."

What do you mean by "quality"? I asked. It sounds a little fuzzy, not very scientific. "I'm talking about the organism as the cause and effect of itself, its own intrinsic order and organization," Brian replied. "Natural selection isn't the *cause* of organisms. Genes don't *cause* organisms. There are no *causes* of organisms. Organisms are self-causing agencies." Now that does sound mystical, I said. "Not if you think in terms of the emergent features of self-organization and the developmental processes we talked about earlier. Not if you think of organisms as the result of a biological attractor, your whirlpool in the sea of a complex dynamical system. When you begin to think of it like that, you begin to approach what I mean by quality."

It still seems to have a tinge of vitalism to it, I suggested. "I don't deny there's a sense of mystery to life," said Brian. "There always will be. But you have to get rid of the idea that there's something added from the outside that is responsible for life. That's the old vitalism. There's nothing added from the outside, it all flows from the inside, from the organism itself, the biological attractor. In my kind of vitalism there's no room for any external mystical 'something' being the cause of it all." You would describe your view as holistic, wouldn't you? "Yes I would. People don't like the word, because it sounds too much like vitalism of the old kind. But it's difficult to get away from the word. I've tried 'integrated' and 'integral,' but I always come back to holism. It works for me."

I had talked to William Provine about the new science of Complexity, with its emphasis on emergence, and its possible role as harbinger of a new push toward a holistic view of nature. Will, a historian of science at Cornell University, was quick with criticism. "The emergentists can claim to be complete materialists and at the same time get out exactly what the vitalists wanted most," he told me. "That is, irreducible lovely properties of evolution going higher and higher, getting more and more complex." But, I said, the Santa Fe folk talk about self-organization in complex systems, about the emergence of patterns in evolutionary models that mimic patterns in nature. They're suggesting that living systems, as complex dynamical systems, are driven to these same patterns. They're saying there is a deep theory to the order we see in nature.

Will remained unimpressed. "I see people trying to make connections between the patterns in the biotic and abiotic worlds, and I'm just not convinced on the face of it," he said. "Tell me what the mechanisms are that produce these patterns, then perhaps I'll get interested." His principal point, as a historian, was that the Santa Fe folk, as the new emergentists, are following a path well trodden. "Each new group of emergentists claims to be more mechanistic than the last," said Will. "It's in the long tradition of a search for purpose in life, a search for the meaning of life. Teilhard de Chardin did it his way. Dobzhansky did it his way. Waddington did it his way. And the Santa Fe people are doing it theirs. In their line of argument, pretty soon you're into free will and determinism."

Is that what you're doing? I asked Brian. Are you looking for the meaning of life, as Will Provine suggests? "He's right, in that people who are studying complex systems are rediscovering the properties the vitalists intuited," said Brian. "There is a convergence of sorts. But, no, we see different things. The vitalists saw an outside force directing life while we see internal, self-organizing principles. So, no, we're not looking for the meaning *of* life, more the meaning *in* life, the generation of order, the generation of pattern, the quality of the organism."

When I put this same question to Stu Kauffman, he was emphatic: "No, I'm not looking for the meaning of life. I'm looking for a deep theory of order in life across the entire spectrum, from the origin of life itself, through the dynamics of evolution and ecosystems, through complexity in human society, and, yes, on a global scale, that of Gaia. I believe the science of Complexity will move us toward that understanding." Can't that be seen as an urge for more than simply an explanation of biological form and order; more a wish that there is some kind of purpose in life? After all, discussions of consciousness often finish up with a wish for something more, a wish for something deep and inexplicable, and this seems to be a human characteristic. "It may sound like that, but language plays tricks on us. As Brian says, and I think he's right, pure Darwinism leaves you without an explanation of the generation of biological form. In the Darwinian view, organisms are just cobbled-together products of random mutation and natural selection, mindlessly following adaptation first in one direction, then the other. I find that deeply unsatisfying, and I don't think that's because I want there to be some purpose in evolution."

You've told me many times that from early in your career you were convinced there must be something deep about the source of order in nature, I said. You wanted to find that source of self-organization, and you did, with your random Boolean networks. And the science of Complexity proclaims it to be true quite generally in the world. And yet no one can say exactly *how* the order emerges, only that it seems to in your model systems. There's still a leap of faith, isn't there, that all this applies to the real world? "You think that deep down I'm looking for a source of order in nature as a psychic solace, the reassuring hand of God on the controls of life?" responded Stu. "We all grew up learning the Second Law of Thermodynamics, which says that systems tend toward disorder. The Second Law is fine as far as it goes, but it turns out to be inadequate as a description of all systems: some systems tend toward order, not disorder, and that's one of the big discoveries of the science of Complexity. So, no, I don't think God has his hands on the controls of life. Let me tell you why some people think that way.

"It has to do with the different way physicists and biologists view the world. Physicists are very comfortable with the notion of self-organization. They see it everywhere. Think of the wonderfully complex patterns of a snowflake, order literally crystallizing out of chaos. But biologists view self-organization with deep suspicion, and it's not difficult to see why. The Darwinian revolution was all about removing seemingly mystical explanations of biological order." William Paley's watchmaker? I ventured. "That's right," said Stu. "Paley's Natural Theology explained biological form as the work of God's hand. Darwin came along and said, No, biological form is the consequence of natural selection. Modern biologists tend to view any suggestion of self-organization as a lurch back toward Paley, so they resist it."

You would like to reformulate Darwinian theory, to include self-organization, is that it? "That's it exactly," said Stu. "We have no theory in chemistry, physics, biology, or beyond that marries self-organization and selection. To do so, as I think we must, brings a new view of life." It extends self-organization from the realm of physics, where it's accepted, into biology, where it is still viewed as mystical at best and heretical at worst? "It does, and it brings us closer to a physics of biology. As Brian says, the science of Complexity will make biological order more intelligible."

By this time in my exploration of Complexity, I had, I have to admit, become something of an enthusiast, although not sufficiently so to satisfy Stu Kauffman's proselytizing passion. "But, aren't you going to proclaim in your book that the revolution is upon us?" he asked incredulously, when one day I explained my position. You may think the revolution is here, I said, but I'm not sure. If everything you say about Complexity is correct, then, yes, we are on the brink of revolution. But you can't say that everything *is* correct, can you? "No I can't," he conceded, "but there's an awful lot of very rigorous science coming out of this. And," he added, "I have a very strong intuition it will turn out to be correct. Intuition is important in science."

My caution stemmed from several sources. From my gut I respond positively to the phenomenon of emergent structures from complex systems—it has the right "feel" to it somehow. Nevertheless, I'm nervous when I can't see exactly how the order is assembled. When Stuart Pimm said to me, "I'm suspicious of emergent properties I can't understand," it struck a chord. Perhaps Stuart and I are too cautious. Will Provine clearly feels the same, and even more strongly. Show us the workings of the machine, and we will become believers, we seem to be saying.

I also came across some downright negative assessments of the Santa Fe Institute's venture. For instance, Oxford University ecologist Robert May told me that what the institute does is "mathematically interesting but biologically trivial." The computer models are too far removed from real biology for his taste, and are irretrievably simplistic. "Well, Bob *would* say that, wouldn't he," was one rebuttal I heard in Santa Fe. Bob has a reputation for arrogance as well as brilliance. "I don't think Bob really knows what's going on here," Stu told me. "If he did, I think he'd see things differently."

Bob did concede that the institute is crammed with talent, and then said that one of things it seemed most talented at was generating hyperbole. Jack Cowan, the University of Chicago mathematician who gave Stu Kauffman his first faculty position back in 1969, agreed. "Don't get me wrong," he said, "there's a lot of good work at the institute, but I often come away from there wondering where some of it is leading." Jack, a member of the institute's science board, has long experience in research on complex dynamical systems. "There have been episodes of tremendous progress in understanding complex systems, but there have also been episodes of unbounded hype," he told me. "Remember Catastrophe Theory?"

In the late 1960s, French mathematician René Thom developed what was, and still is, regarded as an elegant and powerful theory that describes the dynamics of certain nonlinear systems. Specifically, the theory seems able to predict how systems might switch catastrophically from one state to another, hence its name. "There is absolutely nothing wrong with Catastrophe Theory," explained Jack, "except that some of its advocates, including Thom himself, proclaimed that it was virtually a universal law that would explain everything from embryological development to social revolution. Waddington loved it because he thought it might help illuminate embryological development."

Sounds familiar, I said, thinking of the claims made for Complexity. "Doesn't it," Jack replied. "There may well be some universal truths in the theory of Complexity, but the model still needs to be formulated with physics and biology in mind to do it properly. So far, that's lacking." Are you saying Complexity is destined for the same fate as Catastrophe Theory, that it will turn out to be of interest only to a small corner of the mathematical community, with none of its proclaimed wider relevance? "No," Jack replied cautiously. "I'm saying that Complexity theory looks promising, that it may deliver everything its enthusiasts claim for it, but we simply don't know. It's hard to pin down." By "hard to pin down" Jack meant that the mechanics of the different systems—from cellular automata, to embryological development, to ecosystems, to complex societies, to Gaia—have yet to be discerned fully; and when they are, there's a question of how much they will have in common.

So, I asked Stu Kauffman, what of the cautionary lesson of Catastrophe Theory? "It's a beautiful theory," he replied, "and it works perfectly for describing flows on potential surfaces. But most things in nature are not flows on potential surfaces." What makes you think that the theory of Complexity will be any more widely applicable? I asked. "We know that most of nature is composed of nonlinear complex systems, right? And we know that some of those systems, even though they can be described by simple equations, diverge dramatically." You mean, they go chaotic? "Yes, that's chaos theory, just one part of the theory of complex systems. Another part, a much larger part in all probability, describes systems that don't diverge, but instead produce convergent flow, produce structure. This applies in our evolutionary models in computers and in biological systems. True, we don't know precisely *how* the structure is produced, but we do know that it *is* produced, and in a wide range of systems." In other words, you can already say that the theory of complex adaptive systems—what I've been calling Complexity—is widely applicable? "Yes we can; we've demonstrated it." More widely applicable than Catastrophe Theory turned out to be? "There's no doubt in my mind about that."

Language, as Stu said, plays tricks on us, sometimes revealing insights into what is being spoken about and sometimes into the mind of the speaker. I had noticed many times how, when people talked about the dynamics of complex systems, they used the language of purpose, of goal-seeking behavior. "A coevolving system *gets itself* to the edge of chaos," for instance. And, the edge of chaos is a "*favored place to be*," because "that's where computation is maximized" or because "the system optimizes sustained fitness there." Even the phrase "order out of chaos" has a certain numinous quality to it. I wondered whether Will Provine had been correct, suggesting that the people at the Santa Fe Institute really were seeking the meaning of life, and the clue to their underlying motive was revealed in their language. Or perhaps the dynamics of complex adaptive systems are so powerful, so immanently creative, that the language of purposefulness is hard to avoid.

"You're right to say that we sometimes talk like that, as if we are vitalists or something," conceded Chris Langton. "But I think it says more about the nature of the systems we're dealing with than about any hidden motives we might have." The image of the edge of chaos, with its frisson of the unknown, was particularly powerful, we agreed. "It often reminds me of when I was learning to scuba dive in Puerto Rico," said Chris. At the time, in the early 1970s, Chris was helping at a primate colony in Puerto Rico, after he'd left Massachusetts General Hospital in Boston, and before he embarked on his adventure with Artificial Life. He had plenty of time for exploration.

"At first we'd swim in crystal-clear water, and we'd think, Hey, this is deep water here—we're *real* divers. Then one day the instructor took us to the edge of the continental shelf, about a mile off Puerto Rico. As we approached it we could see light blue and then suddenly dark blue, a dramatic dividing line. We were in about sixty feet of water, and we swam to the edge, and looked over. It just dropped away, a slope of about eighty degrees, and you could see that the slope was teeming with life, finally disappearing into darkness. The image stuck: life thriving at the edge, and I've thought about it many times since, kind of iconic for the creativity at the edge of chaos."

Chris's image was indeed powerful. And it turns out to be more than mere iconography, because there is good evidence that evolution is particularly innovative in such waters, poised between the chaos of the near shore and the frigid stability of the deep ocean. Here, at any rate, the abstract edge of chaos and the physical edge meld as one, creative as image and as reality.

Bob May described much of the Santa Fe Institute's attempts at modelling as "biologically trivial." But if the concept of the edge of chaos does indeed translate from computer models to the real world, as Stu Kauffman, Chris Langton, and others firmly believe it will, then there will be nothing trivial about it at all. Stu's coevolutionary model systems get themselves to the edge of chaos, and so, too, do Stuart Pimm's and Jim Drake's ecological models. No one can say yet whether individual ecosystems do the same thing, but the data from mass extinctions at least suggest that, globally, they do. "That's a powerful message of a powerful intrinsic dynamic," said Chris. "Systems poised at the edge of chaos achieve exquisite control, and I believe you see that right the way up to Gaia."

If it's true that, for instance, ecological communities move toward the edge of chaos, where novel properties emerge (such as foodwebs and the ability of a long-established community to resist invasion by alien species), then it seems legitimate to talk about such communities as real systems. It may even be legitimate to think of them as behaving and evolving as a whole, analogous with the superorganism concept that Ed Wilson talked about in connection with social insect colonies. Coevolving communities act in concert as a result of the dynamics of the system; they do so as a result of individuals within the community myopically optimizing their own ends and not as collective agreement toward a common goal; and the communities really do come to know their world in a way that was quite unpredictable before the science of Complexity began to illuminate that world. If true.

"It has to be true," Chris said. "You can see it clearly in our computer models, and it has the right feel for biological systems, too. So, yes, this sense of evolving systems responding to internal dynamics in ways you couldn't predict gives us a very different view of the world." I was getting a sense of biological systems, from the lowest to the highest level in the hierarchy, as behaving like superorganisms, with Gaia as the ultimate. A common sense of the dynamics of life—of living systems—pulses through all levels. I recognized the danger of this line of thinking, and could see how perilously close to mysticism I was slipping. "You can see why we use language the way we do," said Chris.

We were sitting at the round table in Chris's kitchen, and he found more paper on which to draw. "I see a nice rapprochement between mechanism and vitalism in all this," Chris explained as he began to draw. Once again, the image of emergent structure from the interaction of entities in a complex system took shape on the paper. Arrows shooting upward from the lower, local interaction, reaching a cloud hovering above, labelled "Emergent Global Property"; and big arrows sweeping down from the left and right of the cloud toward the interacting entities below, indicating a flow of influence on their behavior.

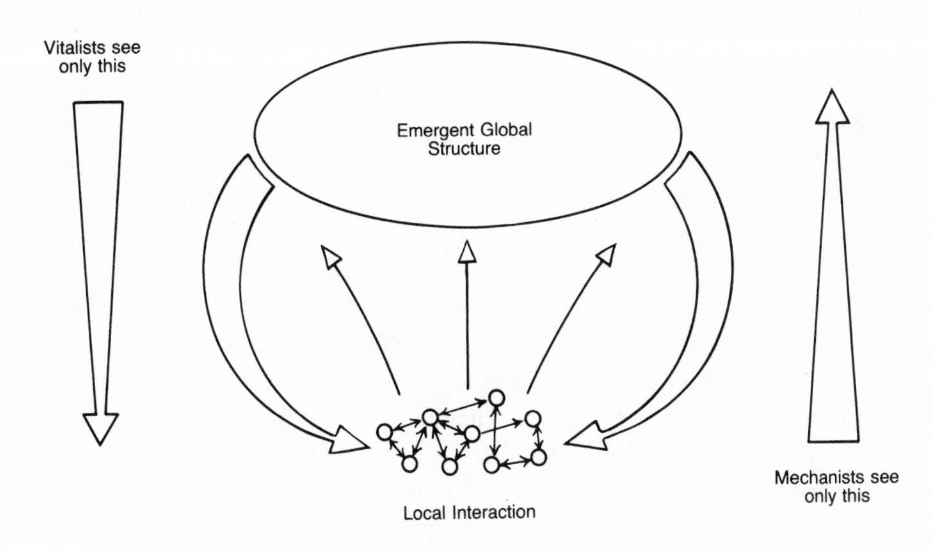

Fig. 10. According to Chris Langton, mechanists and vitalists view the world in opposite ways.

"If you're a strict mechanist, then all you see are the arrows going upward, showing that the local interaction causes some global property, like life or a stable ecosystem," Chris explained. "And if you're a strict vitalist, all you see are the arrows pointing down, indicating some kind of mystical global property that determines the behavior of the entities in the system. Mechanism flows from bottom to top, and vitalism flows from top to bottom," he said with illustrative flourish. "What the science of Complexity gives you is the insight that both directions are important, linked in a tight, never-ending feedback loop. The whole system represents a dynamical pattern, with energy being dissipated through it. The vitalists are going to be disappointed if they look at this kind of pattern as supporting their position, because, take away the energy, and the whole thing collapses. There's nothing external driving the system; the dynamics come from within the system itself."

I could see a coalescence of Platonic and Aristotelian worlds forming here. So, no vital force? I said. "No vital force," said Chris. But you do have something more than the outcome of fundamental mechanical laws in the world. "The old view of the world of nature was that it hovered around simple equilibria. The science of Complexity says that's not true. Biological systems are dynamical, not easily predicted, and are creative in many ways. You've talked to Stuart Pimm; you know that." You said the science of Complexity makes you view the world as *creative*? "Yes. In the old equilibrium worldview, ideas about change were dominated by the action-reaction formula. It was a clockwork world, ultimately predictable in boring ways. In that kind of world, you couldn't have avalanches of extinctions and speciations of all magnitudes driven by the same magnitude of environmental change, for instance, as we see in complex dynamical models."

But biologists have talked about nature as incredibly complex, hardly predictable at all, I protested. "That's true, and there's a paradox here," Chris began. He explained that, yes, nature has been viewed as extremely complicated and difficult to penetrate. The assumption was that this complexity must be the result of complex causes: the action-reaction formula. "The science of Complexity teaches us that the complexity we see in the world is the result of underlying simplicity," said Chris, "and this means two things. First, that you can view the simple systems that underlie it all as being creative, in the way I just mentioned. And second, because simple systems generate complex patterns, we really do have a chance of understanding those patterns. We have a chance of finding simple models that explain the creativity we see. Physicists understand that kind of reasoning, but most biologists believe that simple models can't dissect the complexity that exists out there.

Now we know that they can. We can demonstrate emergence of complexity in computer models, and we are beginning to understand how it will apply in nature."

How far will it apply in nature? I asked. Chris thought for a moment. "You know, I can't see why it shouldn't include the entire spectrum, from embryological development, evolution, the dynamics of ecosystems, complex societies, right up to Gaia—all the things you've been talking about this last year or so." Are we looking at a Theory of Everything here? "I'm not sure you could say it's a theory of everything," he said warily. "I think what we have is an insight into the underlying dynamics of everything. There may be different classes of system within it all, what are called different universality classes." In other words, the overall dynamics of all the systems—from cellular automata through to Gaia—may be common in a general way, but there may be subgroups of systems, the universality classes, that also share detailed dynamics. So, I asked, in principle you might have a general mathematical description for all complex adaptive systems, with more detailed descriptions for each universality class? "Yes, you could." That's *almost* a theory of everything, isn't it? "Almost." And then, with a conspiratorial laugh: "Who knows; maybe one day it will be."

The notion of emergence, so antithetical to much of modern biology, is the principal message of the science of Complexity and its role in illuminating patterns in nature. Emergence of self-organizing dynamics, which, if true, will force a reformulation of Darwinian theory. Emergence of a creativity in the dynamics of complex systems in nature, which, if true, will force a reassessment of the way complexity arises. Emergence of control within ecosystems, which, if true, implies the existence of an "invisible hand" that brings stability from the lowest to the highest level in the ecological hierarchy, culminating in Gaia herself. And the emergence of an inexorable drive toward ever greater complexity and ever greater information processing in nature, which, if true, suggests the evolution of an intelligence sufficiently powerful to contemplate it all was inevitable. Life, at all its levels, is not one damn thing after another, but the result of a common, fundamental, internal dynamic. If true.

I asked Norman Packard how our view of the world would be altered by these implications of the new science of Complexity, if true. He thought for a while, and then in a succinct phrase captured the message at its most appealing:

"We would see the world as having more unity."

We were quiet for a while, Patty Crown, Chip Wills, Jeff Dean, and I, as we sat amid the partly excavated remains of Pueblo Alto. We were looking south, toward Chaco Canyon, which was obscured by a rise in the ground, and out over South Gap, which separates West Mesa and South Mesa. The sun was at its fall high, and a steady breeze hushed through the scattered sagebrush and whisked around the tiny, open rooms behind us. Pueblo Alto was one of the last of the great houses to be built, wasn't it, Jeff? I asked. "It's difficult to get good dates," he replied, "but, yes, it was late." The main part of the structure runs some 350 feet in an east-west direction, with north-to-south extensions running about 150 feet at either end, forming an incomplete rectangle. Each of the three edges of the rectangle was several rooms deep, as in all Chacoan Great Houses, with a few large kivas incorporated here and there. "The latest building was there, maybe A.D. 1130, 1140," Jeff said, sweeping his hand to show how the rectangle was completed, but with a bow-shaped southern wall, thus enclosing a great plaza.

The building at Pueblo Alto was not an isolated activity among the Chaco Canyon Great Houses. If not exactly a frenzy of new construction, then certainly a new commitment to building burgeoned in the period approaching A.D. 1150, and work at Pueblo Alto was part of that. To the innocent eye, such activity might be taken as a sign of vitality in the community. But to the archaeologist experienced in the dynamics of complex societies, it can betoken something more sinister: impending collapse. Joseph Tainter, an archaeologist whom we all had met at the Santa Fe Institute conference a year earlier, has identified several telltale features in the collapse of complex societies. A flurry of collective activity, often involving construction, just prior to collapse is one of them, as if the society was desperately trying to counter rising stress of some nature. Tainter detects the phenomenon in the terminal stages of societies as different as the Roman Empire, the Mayan civilization, and at Chaco.

Is that how it looks to you? I asked Jeff. "I'd say it's a common pattern in the Southwest," he answered. "You often see aggregation of communities, lots of new activity, right before collapse." In our discussion group at the Santa Fe Institute conference, we had talked about some of the patterns in the history of complex societies, the trajectory of evolution through band, tribe, chiefdom, and finally to state. Acknowledging that these terms had to be used loosely, the archaeologists nevertheless agreed that transitions between these different levels of organization—increasing levels of complexity—occurred rapidly. They were punctuations in the history of societies, rapid transitions such as you see in biological systems and in physical systems, too, where they are known as phase transitions.

The recognition of common dynamical patterns in the realms of physics, biology, and society had been important in propelling my exploration of the broad implications of the new science of Complexity. I told Jeff, Patty, and Chip about another behavior pattern that was common to the evolution of complex societies, which I'd learned from University of Michigan archaeologist Henry Wright. The pre-state phase—the chiefdom—can be stable for long periods, Wright had told me. But the transition to the state level of organization was always preceded by a mini-collapse. It was as if the stability of the pre-state phase had to be disturbed before further complexity could be achieved, a process that then happened very fast.

"I was thinking about that kind of thing the other day," Jeff said. "I was thinking about how a community comes to a poised state, when the population levels, resources, and institutional organization reach some kind of stable dynamic. And you can imagine it staying like that until something comes along and disrupts it.

Sometimes you then get an increase in complexity, to state formation, sometimes it collapses to a lower level." Like Chaco, I said, collapsing. "Yes. But, who knows, if conditions had been different—resource possibilities, transport possibilities, that kind of thing—Chaco might have gone on to become a full state instead of collapsing."

The phenomenon of cultural collapse grabs our attention, stirs our emotions. It is, as Jo Tainter says, a reminder of the fragility of civilization. We ask ourselves, What causes such catastrophes? And, Can it happen again? The history of civilization, this brief five-thousand-year episode in *Homo sapiens*'s hundred-thousand-year tenure, is clear: states rise and then they fall, as if marching to the drumbeat of an inexorable dynamic. Proximal reasons for collapse in each case may be very different, such as depleted resources or military conflict, but the overall pattern is unbroken.

The Chaco Canyon community collapsed sometime between A.D. 1150 and 1200, for reasons that are still obscure. "There was a severe drought between A.D. 1130 and 1180," said Jeff, "and the water table fell at the same time, for other reasons. Summer farming would have been tough, no doubt about that." The Anasazi had survived droughts earlier; nothing as intense as this one, it's true. Perhaps the community had reached the point in its trajectory of economic evolution that made it more vulnerable to stress of this kind, I said. "Maybe," acknowledged Patty. "But maybe the perturbation happened elsewhere, the result of other people changing what they were doing. That may have altered the Chacoans' fitness landscape, and made their strategy less successful." Patty's image had been inspired by Stu Kauffman's model of coupled fitness landscapes. "It's the kind of thing that archaeologists have trouble thinking about, but we have to try."

It's often tempting to think about states in history existing and operating in isolation, attracting our attention like clear signals among the archaeological noise. But that's an illusion, and the Chaco system helps us avoid it. Chaco Canyon, with its extraordinary system of roads across thousands of square miles of territory, is seen as a core community of Great Houses, like the spider at the center of a web of influence that encompasses hundreds of smaller settlements. One of them, Mesa Verde, eighty miles almost due north of the canyon, was one such location. And when Chaco Canyon for whatever reason suffered collapse, Mesa Verde took over as the center of influence, albeit on a lesser scale.

"There must have been a lot of contact between Chaco Canyon and Mesa Verde before A.D. 1150," said Chip. "You can see that in the similar styles of ceramics and architecture. And when Chaco lost the influence it once had, the center of power moved north, to Mesa Verde." Architecturally spectacular like Chaco Canyon, the Mesa Verde settlements differed in commonly being built into the face of steep cliffs, the largest of which has been called Cliff Palace. "For almost a century, the Mesa Verde community thrived, just as Chaco had," said Chip. "And then it, too, collapsed. So, you had a virtual repeat of what happened at Chaco." The same kind of dynamics? I asked. "Similar enough to make you think you're looking at the same kind of fundamental processes." History does repeat itself, I thought, and for good reason. "We'd better be on our way," said Chip.

We retraced our steps toward the canyon, following the route of the ancient Anasazi road in some places. With the extent of the sphere of influence of the ancient Chaco Canyon community visible to us from horizon to horizon, I thought about the repeated pattern of the rise and fall of states through history. This was in October 1991, just a few months after the failed coup in the Soviet Union and the brink of collapse of that once great power. George Bush had proclaimed the events in Eastern Europe, of which the disintegration of the USSR became part, as ushering in a "new world order." I remembered a conversation with Chris Langton, animated as always, in which he pulled out a copy of the results of a computer evolution model. "Look," he'd said. "You can see these two species coexisting in a long period of stability; then one of them drops out and all hell breaks loose. Tremendous instability. That's the Soviet Union," he'd said, pointing to the species that dropped out. "I'm no fan of the Cold War, but my bet is that we're going to see a lot of instability in the real world now it's over. That is, if these models of ours have any validity at all."

As we neared the rim of the canyon we began to see the winding course of the Chaco River, the brilliant yellow of the cottonwoods, the earth hues of the ancient sandstone in the steep cliff face, and Hosta Butte in the far distance. Before long we were standing on the edge, once more looking down at Pueblo Bonito, the shell of a community that a historical event pushed into collapse rather than to new heights of complexity. After a few moments Chip said, "Ready?" And soon we were carefully picking our way down the steep, narrow path to the canyon floor.

A SELECTED BIBLIOGRAPHY

Chapter One: The View from Chaco Canyon

A quick introduction to the archaeology of Chaco Canyon can be had in a *Scientific American* article, "The Chaco Canyon Community," published in July 1988, by Stephen H. Lekson, Thomas C. Windes, John R. Stein, and W. James Justice. For nonexpert enthusiasts, Kendrick Frazier's *People of Chaco Canyon*, published by Norton, 1986, is a wonderfully engaging and informative read. (Nothing beats a visit to the place itself, however!)

Heinz R. Pagels's *Dreams of Reason*, Bantam Books, 1989, gives a glimpse of the beginnings of the science of Complexity, and a taste of a keen and imaginative intellect. The proceedings of the first major conference at the Santa Fe Institute, published as *Emerging Syntheses in Science*, edited by David Pines and published by Addison-Wesley, 1988, provides an idea of the scope of Complexity in its formative stages.

Chapter Two: Beyond Order and Magic

Stuart Kauffman gives a short introduction to what he calls "order for free" in a *Scientific American* article, "Antichaos and Adaptation," August 1991. He also has a magnum opus on the topic, *The Origins of Order*, published by Oxford University Press, 1992, which is only for the dedicated. You can reach some of Brian Goodwin's scientific ideas in "Development as a Robust Natural Process," in *Thinking About Biology*, edited by F. Varela and W. Stein, published by Addison-Wesley, 1992; and his more philosophical side in "A Science of Qualities," in *Causality in Modern Science*, edited by Willis Harman, in press. For a view from the opposition, Richard Dawkins's *The Blind Watchmaker*, Norton, 1987, is unbeatable.

Chapter Three: Edge of Chaos Discovered

The first papers on the topic were Chris Langton's, "Studying Artificial Life with Cellular Automata," *Physica* 22D (1986), 120–49, and Norman Packard's "Adaptation Toward the Edge of Chaos," Technical Report, Center for Complex Systems Research, University of Illinois, CCSR-88-5 (1988). Not easy reads, but there's nothing else in the popular press yet. However, Per Bak and Kan Chen give an account of self-organized criticality in an article with that title in *Scientific American*, January 1991.

Chapter Four: Explosions and Extinctions

The most accessible, detailed account of the Cambrian explosion and new interpretations of the processes behind it is to be had in Stephen Jay Gould's *Wonderful Life*, Norton, 1989. Stuart Kauffman gives his alternative explanation in "Cambrian Explosion and Permian Quiescence: Implications of Rugged Fitness Landscapes," *Evolutionary Ecology* (1989), vol. 3, 274–281.

For an exploration of ideas on the causes of mass extinction there is nothing better than David Raup's superbly clear and readable *Extinction: Bad Genes or Bad Luck?*, Norton, 1991.

Chapter Five: Life in a Computer

Steven Levy's recently published *Artificial Life* (Pantheon, 1992) is a good narrative jog through the issues in and personalities of the topic. Serious students will want to immerse themselves in the proceedings of the two workshops on artificial life, edited by Chris Langton and friends: *Artificial Life* and *Artificial Life II*, Addison-Wesley, 1989 and 1992.

Chapter Six: Stability and the Reality of Gaia

There are scores of books purtporting to be about the Gaia theory, but it's best to stick with Lovelock's own, the most recent of which is *The Ages of Gaia*, Bantam, 1990. For a mélange of supportive and critical voices, *Scientists on Gaia*, edited by Stephen H. Schneider and Penelope J. Boston, published by MIT Press, 1991, presents the proceedings of a conference on Gaia, organized by the American Geophysical Union in San Diego, 1988.

In *The Balance of Nature*, University of Chicago Press, 1991, Stuart Pimm gives a glimpse of ecological thinking of the next decade. Not meant as a general book in any sense, it nevertheless is a compelling argument, beautifully presented.

Chapter Seven: Complexity and the Reality of Progress

The small volume *Evolutionary Progress*, edited by Matthew Nitecki, published by the University of Chicago Press, 1988, is the best immersion in the issues. The published proceedings of a conference of the same name at Chicago's Field Museum in 1987, the book conveys the overwhelming message that there is no progress in evolution. Robert Richards's *The Meaning of Evolution*, University of Chicago Press, 1992, is a thoughtful and eloquent exploration of some of the history of the ideas and the reality of Charles Darwin's position.

Daniel McShea's paper "Complexity and Evolution: What Everybody Knows," *Biology and Philosophy* (1991), vol. 6, 303–24, is an excellent overview of what everybody doesn't know. Those who wish to read Herbert Spencer surely know where to find him.

Chapter Eight: The Veil of Consciousness

So much has been written about consciousness, it's hard to know where to start. The three major books of late are Daniel Dennett's *Consciousness Explained*, Little, Brown, 1991; Nicholas Humphrey's *A History of the Mind*, Chatto and Windus, 1992; and Roger Penrose's *The Emperor's New Mind* Oxford University Press, 1989. Although (so far) the recipient of the least notice of the three, Humphrey's book is surely the best, as well as being crafted in the most elegant prose. Colin McGinn's article "Can We Solve the Mind-Body Problem?," *Mind*, (April 1989), vol. 98, no. 390, is an engaging argument over the possible inexplicability of consciousness. Two reviews of the topic (in the guise of reviews of recent books) are worth tracking down: "What Can't the Computer Do?," by John Maynard Smith, in the 15 March 1990 issue of the *New York Review of Books*, and "A Parliament of Mind," by Adina L. Roskies and Charles C. Wood, in *The Sciences*, May/June 1992.

On a slightly different tack, Patricia Churchland and Terrence Sejnowski's *The Computational Brain*, MIT Press, 1992, is a masterwork on the neurobiology underlying the generation of mind. The January 1990 issue of *Scientific American* contains a useful debate over artificial intelligence, with John Searle on one side and Paul M. Churchland and Patricia Churchland on the other.

Chapter Nine: The View from the Edge

The is no more complete work on ants than Edward O. Wilson and Bert Hölldobler's *The Ant*, Harvard University Press, 1991.

Joseph Tainter's scholarly but highly readable book, *The Collapse of Complex Societies*, Cambridge University Press, 1988, is a fascinating account of the repeated pattern of cultural collapse throughout history.

And if anyone still believes that complex societies may not be toppled from a poised, quasi-stable condition into sudden chaos, they should start reading the newspapers.

I N D E X

A

Acetabularia acetabulum (mermaid's cap), 36–39
"Adaptation Toward the Edge of Chaos" (Packard), 54
Alice Through the Looking Glass (Carroll), 58
Alvarez, Luis, 76
American Museum of Natural History, 100, 143
Anasazi Indians, 2–8, 17, 18, 70, 71, 151, 194
Anderson, Philip, 60, 169
Andrews, Roy Champman, 143
Ant, The (Hölldobler and Wilson), 172
Aristotle, 24
Aristotelianism, 178, 189
Arizona, University of, 2, 17, 47
Arizona State University, 2
Arms races, 148–49
Artificial intelligence, 155, 156, 159–64
Artificial life, 47–48, 53, 87–104, 150, 171, 187
Artificial Life (Langton), 90
"Artificial Life: an Ecological Approach" (Ray), 91
Asteroid impacts, 76–77, 81
Attractors, 20–21, 27, 46
  morphogenic, 39, 40, 72–74
Ayala, Francisco, 137–38, 169–70

B

Bak, Per, 60–61, 77, 81
Balance of Nature, The (Pimm), 119
Bateson, William, 40
Behavioral and Brain Sciences (Harnad), 154
Biomorphs, 102
Blind Watchmaker, The (Dawkins), 34, 102, 148
Bohr, Niels, 178–79
Bonner, John Tyler, 135
Boole, George, 27
Boolean networks, 27–32, 35, 38, 41–42, 44–46, 55, 121, 130, 132, 183
  fitness landscapes and, 57, 125
Boston University, 47, 114
Brain, 13–14
  consciousness and, 150, 153–64, 167–68
  evolution of, 145–46
Broken stick model, 80
Brookhaven National Laboratory, 60
Broom, Robert, 141
Buffon, Georges de, 140
Burgess Shale, 66–67, 70

"Burgess Shale Faunas and the Cambrian Explosion" (Morris), 19
Bush, George, 195
Butterfly effect, 11

C

California, University of
  Berkeley, 26, 36, 64, 161, 166
  San Diego, 123, 161
  San Francisco, 23
  Santa Barbara, 19, 64, 76
  Santa Cruz, 52
California Institute of Technology (Cal Tech), 9, 48
Calvin, William, 153
Cambrian explosion, 17–19, 21, 30, 63–74, 83, 102–3
Cambridge Go Club, 88
Cambridge University, 19, 66, 153, 157
Capitalist economy, 13, 59
Cartesian dualism, 154, 167
Case, Ted, 123
Catastrophe Theory, 75, 76, 185, 186
Cellular automata, 46–50, 53–56, 90, 97, 104, 137, 185, 191
Chaco Canyon culture, 1–10, 20–22, 70, 151, 192–96
Chaos (Gleick), 12, 91
Chaos theory, 12, 46, 52
Chicago, University of, 45, 57, 58, 74, 130, 133, 143, 185
Chomsky, Noam, 168
Churchland, Patricia, 161, 163–67
Churchland, Paul, 161
Cincinnati General Hospital, 45
Circadian rhythms, 23
Cohen, Morel, 130
Cold War, 196
Computational Brain, The (Churchland and Sejnowski), 163
Concept of the Mind, The (Ryle), 154
Connectedness, 81–82
Connection Machine–5, 160
Consciousness, 13–14, 145–46, 149–71
Consciousness Explained (Dennett), 155
Conway, John, 49, 61, 98
Coombe Mill Experimental Station, 107, 108, 110, 128
Cornell University, 57, 181
Cosmology, 133
Cowan, Jack, 185
Cretaceous extinction, 76
Criticality, self-organized, 60–61, 127
Crown, Patricia, 2–4, 7, 8, 16–18, 192–94
Crutchfield, J., 56
Cuvier, Georges, 40, 75

D

Dalhousie University, 115
Dartmouth College, 25
Darwin, Charles, 23, 24, 33–35, 37, 41–43, 66, 74–75, 121, 134, 138, 140, 143–44, 146–48, 183
Dawkins, Richard, 34, 41, 101–2, 114–15, 117, 146, 148–49
Dean, Jeffrey, 2–6, 8, 16, 20, 192–94

Deep Thought, 161
Delaware, University of, 85, 87, 89, 91, 95, 103
Dennett, Daniel, 154–58, 160, 168
Derrida, Bernard, 55, 56
Descartes, René, 154
DNA, 15, 35, 93, 94, 102, 153, 166, 179–80
Dobzhansky, Theodosius, 137, 182
Doolittle, Ford, 115
Drake, Jim, 124–27, 188
Dreams of Reason, The (Pagels), 10
Driesch, Hans, 40
Dualism, 154, 161, 167
Dynamical Systems Collective, 52

E

Eccles, John, 154
Ecole Normale Supérior, 55
Ecological hypothesis, 67
Ecosystems, 13, 83, 120–21, 175, 185
  computer models of, 60, 123–27, 188
  patterns in, 85–86
  perturbations of, 61–62, 77–78, 81
Edinburgh University, 28
Einstein, Albert, 42
Eldredge, Niles, 100
Electron Capture Detector, 108
Elements of Physical Biology, The (Lotka), 116
Embryology, 13, 26, 31, 36, 38–39, 175, 185
  mechanics of, 72
Emperor's New Mind, The (Penrose), 155, 161, 162
Enlightenment, the, 40
Evolution
  biological, 15, 17–20, 31, 35, 57, 86–87, 134–36, 180, 187
  bricolage concept of, 164
  computer models of, 87–96, 99–104, 131, 137, 171, 181, 188
  creativity of, 52
  cultural, 9, 18–21, 47, 70, 71, 193–94
  progress and, 133, 138–48, 169–70
  See also Natural selection
Evolutionary Ecology (Kauffman), 68
Evolution of the Brain (Eccles), 154
Extinctions, mass, 63–65, 67, 69, 74–83, 96, 100, 103–4
Eye, evolution of, 34, 38–40

F

Farmer, Doyne, 52, 53, 91, 92, 97
Field Museum (Chicago), 138, 139, 169
First Artificial Life Conference, 90
Fitness landscapes, 57–59, 61, 104, 109
  rugged, 68, 125–27
Fontana, Walter, 91
Foodworks, 121–22, 127, 175, 188
Forrest, Stephanie, 91

G

Gaia: A New Look at Life on Earth (Lovelock), 114
Gaia hypothesis, 60, 82, 107–9, 112–15, 117–18, 120, 127–29, 175, 182, 186, 188, 191
Game of Life, 49, 61, 90, 98, 133
Gell-Mann, Murray, 9, 14–16, 85
Genetic algorithms, 46, 54
Genetics, 26–27, 29–31, 35–36, 57, 179–80
Genomic theory, 67, 68
Gleick, James, 12, 91
Goethe, Johann Wolfgang von, 40, 41
Golding, William, 113
Goodwin, Brian, 28–29, 32–43, 72–74, 166, 176, 180–82
Gould, Stephen Jay, 43, 66, 72–74, 79, 80, 100, 139–45
Gradualism, 66, 75
Great Chain of Being, 134
Great Smoky Mountain National Park, 120
Guelph, University of, 146
Gumerman, George, 18–19, 70, 71

H

Hang-Kwang Luh, 127
Harnad, Stevan, 153–54
Harvard University, 25, 87, 107
  Museum of Comparative Zoology, 140, 172, 177
  Science Center, 88
Hillis, Danny, 160
History of the Mind, A (Humphrey), 158
Hölldobler, Bert, 172, 173, 175–76
Humphrey, Nicholas, 153, 157–60
Hutton, James, 75
Huxley, Thomas Henry, 140
Hyatt, Alpheus, 143

I

Iron Age, 128

J

Jacob, François, 26, 29, 41, 164
James, William, 118, 169
Jaynes, Julian, 153
Johnsen, Sonke, 77

K

Kant, Immanuel, 25, 40
Kauffman, Elizabeth, 32
Kauffman, Stuart, 16, 19–20, 23–32, 35, 38, 40–47, 55–62, 64, 68–70, 77–79, 81–82, 96, 103–4, 108, 109, 117, 121, 125–28, 130–32, 148, 149, 165, 170, 182–87, 194
Koobi Forest, 16

L

Lamarck, Jean-Baptiste de, 146
Lambda parameter, 50, 53
Langton, Chris, 10–14, 16–18, 20–22, 47–51, 53–56, 87, 90–92, 94, 97–101, 104,

150–52, 170, 187–91, 195–96
Language, 157, 168
Larson, Gary, 156
La Selva Biological Reserve, 84–87, 173
Leibnitz, Jacob, 11
Levin, Simon, 57
Levins, Richard, 131
Lewontin, Richard, 131
Lindgren, Kristen, 99–100, 131, 132, 137, 175
Linnaean Society, 129
Linnaeus, Carolus, 24
"Lopsided Look at Evolution, A" (Kauffman), 64
Los Alamos National Laboratory, 14–15, 52, 53, 90–95, 97, 150
Lotka, Alfred, 116–17
Lotka-Volterra cycle, 95–96, 117
Lovelock, Jim, 107–19, 128–29
Lovelock, Sandy, 107
Lyell, Charles, 75

M

MacArthur Foundation, 24, 85
McCulloch, Warren, 32, 44–45
McGill University, 28
McGinn, Colin, 167–69
McShea, Dan, 132–38, 145–48
Marcus Aurelius, 42
Margulis, Lynn, 114, 115
Mars, life on, 112
Massachusetts General Hospital, 47, 97, 187
Massachusetts Institute of Technology (MIT), 32, 44, 45, 88
Mass extinctions, 63–64, 67, 69, 74–83
  in computer ecosystems, 96, 100, 103–4
Materialism, 154, 167, 168
May, Robert, 117, 184–85, 187
Mayans, 193
Meaning of Evolution, The (Richards), 143
Mechanism, 178, 180, 182, 188–89
Mesopotamia, 17
Michigan, University of, 47, 48, 53, 92, 97, 133, 193
Mind-body problem, 154, 167
Minsky, Marvin, 160
Molecular biology, 36, 179
Monod, Jacques, 26, 29
Montefiore, Hugh, 108
Morphology, 36, 37, 39, 40, 72–74, 88, 134–36, 175
Morris, Simon Conway, 19, 72
Motion, laws of, 11–12
Multiple personality syndrome, 157

N

National Aeronautical and Space Administration (NASA), 112
National Institute for Medical Research, 107
National Institutes of Health, 45
Natural selection, 24, 25, 27, 34–35, 40–43, 54, 66, 72, 80, 147, 148, 180, 182
  in computer model, 93
  extinction and, 75
  fitness landscapes and, 57–59

Natural selection (*continued*)
Gaia hypothesis and, 115
progress and, 143, 144
Natural Theology, 33, 41, 183
Nature, 117
Navajo, 5
Nazism, 144
Neo-Darwinism, 40–42
Neural networks, 25, 46, 164, 166–67
Neurophilosophy (Churchland), 163
Neuroscience, 163–66
New Age, 5, 128
New Mexico, University of, 2
Newton, Isaac, 11
New York Times, 141, 172

O

Olduvai Gorge, 16
Open University, 32
Origins of Order, The (Kauffman), 56
Origin of Species (Darwin), 34, 66, 75, 87, 121, 144, 148
Osborn, Henry Fairfield, 143
Owen, Richard, 40
Ownby, John, 121
Oxford University, 25, 26, 28, 101–2, 117, 153, 161, 184

P

Packard, Norman, 51–56, 137–38, 145, 146, 148, 169–71, 192
Pagels, Heinz, 10
Paley, William, 33–34, 183
Pangea, 82
Parallel-processing computer, 161, 162, 164
Pennsylvania, University of, 16, 23, 24, 44, 45, 77, 131
Penrose, Roger, 153, 155, 161–63, 168
Permian extinction, 65, 67, 69, 79, 82
Perturbations, 61–62, 69, 127
extinctions and, 77–78, 81
Phase transitions, 17, 20, 51, 54, 55
Phyla, 65, 67, 71
Pimm, Stuart, 119–28, 184, 188, 189
Plate tectonics, 82
Plato, 133, 162
Platonism, 178, 189
Pliny the Younger, 130
Post, Mac, 123, 124
Postgate, John, 114
Poundstone, William, 133
Power law distribution, 50, 77–81, 103–4, 127
Pre-Columbian societies, 2
Prediction Company, 52, 169
Princeton University, 48, 52, 53, 60, 135, 153
Institute for Advanced Study, 48, 52
Principles of Geology (Lyell), 75
Principles of Psychology (James), 169
Prisoner's Dilemma, 99–100, 132
Progress, evolutionary, 133, 138–48, 169–70
Provine, William, 181, 184, 186
Punctuated equilibrium, 20, 21, 100–101, 119
Purdue University, 124

Q

Qualia, 155, 157
Quantum physics, 179

R

Racism, 141–43
Rassmussen, Steen, 91
Rational Morphology, 40–41
Raup, David, 74–82, 104, 133
Ray, Tom, 84–97, 99–105, 123, 131, 137, 150, 173, 175
Recursive Universe, The (Poundstone), 133
Red Queen effect, 58, 59
Reductionism, 35, 179, 180
Remote sensing techniques, 6
Richards, Robert, 143–44
Rockefeller Foundation, 42, 130
Rockefeller University, 10
Rockford College, 47
Roman Empire, 193
Royal Society of London, 113, 129
Rugged landscapes, 64, 68, 125–27
Ruse, Michael, 146
Rutgers University, 167
Ryle, Gilbert, 154

S

Saint-Hilaire, Goeffroy, 40
Santa Fe Institute, 9–21, 24, 45, 47, 48, 70, 87, 99, 103, 132, 137, 140, 147, 155–57, 169, 181, 182, 186, 187, 192
Science, 19, 36, 64, 70
Scientific American, 161
Searle, John, 161
Sejnowski, Terrence, 163
Self, sense of, 154, 157
Self-organization, 43, 183, 191
Selverston, Allen, 166
Sepkoski, Jack, 76, 79
Service, Elman, 19
SimAnt, 172
Simberloff, Dan, 79–80
Smith, Adam, 13, 59
Smith, John Maynard, 42–43, 45, 57, 130, 146
Smithsonian Institution, 66, 173
Social Darwinism, 147
Sociobiology (Wilson), 173
Soviet Union, 195
"Speaking for Our Selves" (Dennett and Humphrey), 157
Speciation, 61, 119
Spencer, Herbert, 140, 147–48
Spin glasses, 46
Stanley, Steve, 80
State cycles, 27–28, 30
State formation, 9, 17, 19, 21
Stebbins, G. Ledyard, 137
Stent, Gunther, 36
"Studying Artificial Life with Cellular Automata" (Langton), 53
Sussex, University of, 42, 43

T

Tainter, Joseph, 192–94
Technological innovation, 70–71
Teilhard de Chardin, Pierre de, 141, 182
Teleology, 114, 115
Temporal Organization in Cells (Goodwin), 28, 29, 32

Tennessee, University of, Knoxville, 119
Thermodynamics, Second Law of, 183
Thom, René, 185
Thompson, D'Arcy Wentworth, 40–41
Tierra, 87, 94–96, 99–104
Times Literary Supplement, 155
Tufts University, 154, 155, 157
Turbulence, 15, 101, 166
Turing, Alan, 49, 160–61
Turing Machine, 49
Turing test, 160–61

U

Universal computation, 49, 54, 55
Up From the Ape (Hooton), 142

V

Valentine, James, 64–65, 67, 137
Van Valen, Leigh, 58, 148
Vietnam War, 26, 47, 97
Virtual computers, 92
Vitalism, 24, 178–79, 181, 182, 187–89
Volterra, 117
Von Neuman, John, 46, 48

W

Waddington, Conrad H., 28, 41, 130, 179, 180, 182, 185
Walcott, Charles, 66, 167
Wallace, Alfred Russell, 34, 43
Wall Street Journal, 9, 52
Washington, University of, 153
Watson, Jim, 153
Weather prediction, 11
Weishbuch, Gerard, 55, 56
Wheeler, William Morton, 177
Whittington, Harry, 66, 67
Wills, Chip, 1–8, 16–18, 192, 193, 195, 196
Wilson, Edward O., 87, 136, 138, 146, 172–78, 188
Wittgenstein, Ludwig, 157
Wolfham, Steve, 48–50, 52, 56
Wolpert, Lewis, 131
Wonderful Life (Gould), 66
Woods Hole Marine Biological Laboratory, 79
Wright, Henry, 193
Wright, Sewell, 57

Y

Yoffee, Norman, 17

ABOUT THE AUTHOR

After working on the staff of *New Scientist* in London for nine years and then of *Science* in Washington, D.C., for a further nine, Roger Lewin became a full-time free-lance writer in 1989 in order to concentrate on books, of which *Complexity* is the second. Earlier, in what was laughingly called his "spare time," he wrote about a dozen popular-science books, including three with world-renowned anthropologist Richard Leakey. He likes writing books that require visits to exotic places.

Phoenix Bookstore
Lambertville, NJ
Sat 25 June 1994
$9.00 less 10%